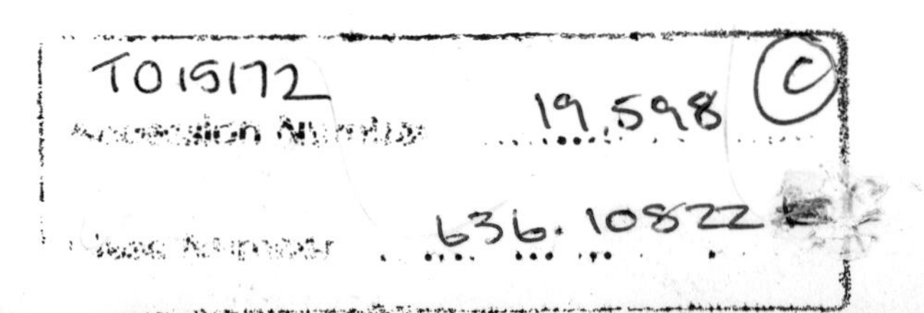
T015172
Accession Number 19,598
Class Number 636.10822

Janet Lorch

FROM FOAL TO FULL-GROWN

David & Charles

FOREWORD

One never stops learning with regard to any aspect of horse breeding and management. It is with this in mind that I have the greatest of pleasure in writing the foreword to this very practical and well illustrated book *From Foal to Full-Grown*.

There are so many questions which need answering when breeding a foal, whether from a 'first-time breeder' or from a professional who has been breeding horses for a lifetime. Foaling is a natural process and rarely causes problems, but it is imperative to know what to do if things go wrong. The aftercare of mare and foal is so important, and this book gives clear, concise details of what to look for and when to call a vet. The foal's first year is an important period of his life; good management in terms of regular worm control, farrier's visits and general handling are all discussed in detail.

My wife and I have been breeding horses for nearly thirty years, and the arrival of a newborn foal still gives us the same amount of pleasure as the very first one.

Choosing the right mare and stallion is of the utmost importance, and this book is an excellent guide; remember that it costs the same amount of money to breed a bad foal as it does to breed a good one!

Should you have some bad luck when breeding a foal, don't give up. We need enthusiastic horse breeders to keep up with the continentals, and this book will help us all to do just that.

John Rose

JOHN ROSE
Broadstone Stud, Oxfordshire, 1993

This book is dedicated to Muschamp Danube, an admirable, proud stallion who has taught me to respect and understand the needs of a good sire

Acknowledgements

The author would like to thank the following people for their assistance and time while the book was being written: staff at the Muschamp Stud, in particular Grainne Blanchfield, Sam Ingray and Gordon and Janice Morgan; stud veterinary surgeon Angus Campbell BVM&S MRCVS of Scott Dunn; Johanna Vardon, The National Foaling Bank; Beverley Holtom; Maggi Marzi, Germany; Kay Millward, Hallagenna Stud; Hilary Vernon of Turf and Travel; Debbie Wallin, Chairman BWBS and typist Alison Freedman.

Photographs by Bob Langrish

A DAVID & CHARLES BOOK

First published 1993
Reprinted 1994 (twice)

A catalogue record for this book is available from the British Library.

ISBN 0 7153 9976 4

Typeset in 11/13.5pt Times and Fenice ITC by ABM Typographics Ltd, Hull, England and printed in England by BPC Paulton Books Ltd. for David & Charles
Brunel House Newton Abbot Devon

CONTENTS

1

SELECTING BROOD MARE AND STALLION

Some fifty years ago when you wanted to breed from your mare and did not actually own a stallion there was little choice available. Transporting a horse was not simple, and facilities enabling you to leave your mare at the home of the stallion were non-existent. Stallions at that time 'walked' their county, or simply ran with their mares. Today the choice of stallions is overwhelming. However, they are often so valuable that covering is done in-hand to reduce the risk of injury, and the mare is taken to the stallion.

In this chapter I will give you some ideas as to where to find the 'right' stallion for your mare, but prior to that, I ask you to think carefully as to whether your mare is a suitable candidate for breeding, and why you want to breed from her. Can you cope with the future – your mare and another one 'in tow'? Will the foal be eligible for registration papers? What will he be worth? Do you have any idea of what to look for when you take your mare to stud?

What constitutes good stud facilities? When is the best time to take your mare to stud? How long will she have to stay away? In fact, what is the whole exercise going to cost? Is it feasible to take your mare to be covered when she is fully in season and then bring her straight back home? This is a method widely practised in Germany and other European countries. All these points will be noted in the first two chapters, together with specific guidelines on what needs to be discussed during stud visits, the condition your mare should be in when she goes to stud, the optimum breeding age, swabs and other related recommended subjects.

The suitable mare

A mare whose own performance has not been tested should at least have a pedigree that reveals known performance bloodlines – though good conformation and performance ability are naturally closely linked, as a horse or pony must usually be well 'put together' in order to have the necessary agility to perform well.

Again, why breed? If you are the proud owner of a family cob or pony mare and your children have either grown out of her or become bored with riding, then this could well be your reason for wanting to put her in foal. Provided you know she is sound and of equable temperament, and you are sure that you have done your homework thoroughly and are ready to enter the commitment, then why not?

However, if your mare is permanently unsound, I suggest you first check with your vet as to the reason why. Many lameness problems can be hereditary and he will advise you if you should breed from her or not: it is a grave mistake knowingly to continue a hereditary problem.

POINTS TO CONSIDER CAREFULLY BEFORE BREEDING FROM YOUR MARE

1 Does she have poor conformation?
2 Does she have an unstable temperament?
3 Is there permanent lameness that is known to be hereditary?
4 Is she too old (p9)?
5 Does she totally lack ability to perform?

Most studs request information on soundness when you complete the nomination form, and reserve the right *not* to accept a mare at stud if she has a known hereditary disease or lameness. Reason: no stud wants to have offspring bred from their stallion which may ultimately have a problem. You must remember that any stallion owner always hopes that his stallion's progeny will do him justice and be a good advertisement for the stud in the future.

In Europe, and more recently in the UK, too, most breeds have a grading system for stallions *and* mares. Both have to be graded, which means they have to be inspected for suitability as a breeding horse, and that no offspring from ungraded stock will receive papers (see p20). This may be ruthless, but it is a certain way of upgrading what is bred.

Think carefully before you breed – after all, if you want to sell the youngster, only the best will sell at a price that will make your investment break even.

The brood mare

Once your mare has given birth to a foal she is known as a 'brood mare', although even before this event you need to decide if you are going to put her back into foal. Of course the final decision will depend on the well-being of the mare and foal after the birth, but provisional plans need to be made beforehand (see p24).

The single mare owner may well have other ideas for her – namely to bring her back into work – and anyway may not have the desire, finance or facilities to continue breeding on an annual basis. However, a mare which has been retired from competition and produces a quality foal, may well have no other potential use (too old, unsound) so careful consideration needs to be given as to her future.

Studs whose livelihoods depend largely on the sale of their offspring, will probably breed from each of their proven brood mares on an annual basis. To a certain extent nature will dictate – if the mare does not conceive easily maybe her body requires a rest from carrying foals.

As regards the domestic pet, it is known that breeding on a continual basis without respite can produce smaller litters, and the newborn can be reduced in size. Research into breeding from the horse has not published matter to this effect, but it is said that *unless very careful feeding of protein and vitamins* is carried out during pregnancy, with special attention to the calcium and phosphorus ratio, the offspring may lack in size and bone (see p52). Certainly it is rare to find a mare in the wild who breeds each and every year for more than seven or eight years.

The right age to breed

The mare is said to be 'at the correct age' to breed from when she is between the ages of **three and twelve**. Although there are no hard and fast scientific facts on this subject. However, if you take a look at nature's way, then **wild mares** normally conceive at two years old and miss out the following year. I personally believe that a two-year-old filly is too young to put in foal. She is not yet mature, and bone formation continues well into the horse's third year: carrying a foal at this critical time in a filly's growth pattern may well be a hindrance. There are, of course, always exceptions – if a two-year-old has been done well, is well grown and has good heart-room plus depth of girth, and if she conceives easily, then let nature take its course. Putting a mare in foal at two is more widely practised in America and on the continent than in the UK. However, to breed from a three-year-old is certainly a good idea; she will mature well during the gestation period, and is thereby given a good period of 'mental freedom'; though I do recommend that she is lightly backed either before or after covering (see p167).

After she is twelve years old, conception can often be more difficult, especially if your mare is a maiden, or unfit/out of condition. **But do not** be led to think that it is easy to get a mare in foal at any age – like women, horses do not always conceive when requested to do so! Nevertheless, mares over twelve can be bred from, there is no doubt, and I have known an eighteen-year-old maiden conceive; though she was a fit horse retiring from competition and much assistance was given by the visiting stud vet. Mares with foals at foot are often easier to get into foal and can go on breeding until their early twenties, *but* this does depend on their overall general health and soundness.

In the final analysis **nature** will always take its own course; nonetheless, the modern stud vet has made a great impact on breeding today, and most studs have access to a specialist stud vet who will visit on a regular basis; the large Thoroughbred stud will have a live-in stud vet who will be on hand for difficult foalings and daily stud practice.

Note: **a barren mare** is a mare that has previously been bred from, but is not currently in foal or with foal at foot.

**SUMMARY:
RECOMMENDED BREEDING AGES**

Maiden mares 3 years old up to 12/14 years old.
Older maiden mares At 14 years and over, these can be more difficult to get in foal. The mare should be **fit** and **not overweight** (see p24). Once over 15 years, their success rate is lowered.
Brood mares *ie* mares with foal at foot, can breed up to about 23 years of age. Once over 14 years of age, it is often more likely that they will conceive if in lactation (*ie* with foal at foot). Much depends on the physical fitness of a particular mare.

Selecting the right stallion for your mare

Having decided that you are going to breed from your mare, you now need to find the most suitable stallion at a price that you can afford. How do you go about this? In many countries there are several different breeds to choose from: in some, such as native ponies, hundreds of years of history can be traced back through their pedigrees and bloodlines.

First: make sure that you know as much as possible about your mare and are familiar with her own particular breeding, pedigree, performance, behaviour and conformation. Make a list of her poorer characteristics for reference – these are the points that you will be seeking to improve, and will have a major influence on the choice of stallion.

Second: make a decision as to what type of horse or pony you want to produce, and in which sphere, potentially, you want him to perform: that is, do you want an all-round weekend riding horse, an eventer, a show jumper, a dressage or a driving horse? Do you want a show pony, a family pony, or a junior competition pony in whatever sphere?

Third: gather together information on stallions that may meet your criteria. In the UK, information can be obtained from various breed societies

A Selle-Français stallion: this picture was used on his stud card so mare owners could study his conformation before deciding to visit

and magazine sources. Having made your selection, request stud cards from the relevant stud. Each **stud card** will note the essential details of the stallion including date of birth, height, bone and bloodlines (see p13), plus information about the stallion and related species.

It is estimated that today, only 1 to 4 per cent of male horses are chosen for breeding purposes. The standard of stallion selection has to be maintained at a very high level, due to the fact that a stallion is able to sire a large number of offspring on an annual basis, as opposed to the mare who is only able to produce one per year. Despite this fact there is still a wide choice of quality stallions at stud, and additional criteria must be borne in mind when making your selection.

All breed societies set a minimum standard for stallions, and in some breeds the grading (that is, the approval of stallions to stand at public stud) is very stringent. There is always a strict veterinary examination which ensures that no graded stallion has any hereditary defects, and in most breeds, an additional judging of conformation, paces, and trueness of type is made; pedigree/bloodlines are also taken into consideration.

Vital, a famous show-jumping stallion. It is sometimes useful for a mare owner to see a picture of a potential sire participating in his particular sport

46

POINTS TO CONSIDER WHEN CHOOSING A STALLION*

1 Breed: is the breed going to enhance the progeny in type and bone?
2 Conformation: does the stallion have points that may better your mare's (for example if your mare has a long back, is the stallion compact to compensate)?
3 Size: does the stallion differ greatly from the height and build of your mare? (Although scientific evidence has proved that a Shetland mare can produce a foal from a Shire **because the size of foal is determined by the mare's body size,** I believe you should not err too far from what nature intended.)
4 Presence: does the stallion show presence, an important trait for the offspring?
5 Paces: are the stallion's paces correct, and does he have the type of movement necessary for the discipline that you intend to breed for?
6 Performance: does the stallion himself have a proven performance record, or alternatively, known performance bloodlines?
7 Temperament: does the stallion have an equable temperament? This is a difficult point to assess when you visit a stud, but a good, kind eye often indicates a pleasant nature.
8 Progeny/offspring: have you heard of the stallion's progeny? Have you seen any of his offspring? Are they the type that you are seeking to breed?
9 Do you personally like the stallion? In the final analysis you must genuinely like the stallion – he must 'take your eye'.
10 Colour: what colour is the stallion known predominantly to produce in his offspring?

*List does not indicate order of priority.

SOURCES OF INFORMATION

1 The relevant breed society, which will have a list of approved stallions.
2 *Directory of Approved Stallions* published by The National Stallion Association (UK).
3 *Hunter Stallion Guide* published by Vardon Enterprises.
4 *The Warmblood Sires Directory*.
5 Advertisements for stallions at stud in the two major stallion issues of *Horse and Hound*, plus other specific issues from December onwards.
6 Advertisements and articles on stallions in other horse journals, many of which have **special stud issues running from February through to the spring**.
7 At the time of writing, a video of approximately fifty stallions at public stud available from The Video Horse Agency.

In European Warmblood breeding a stallion must carry out a **100-day performance test** in his third or fourth year to qualify for grading. He is kept at one particular centre with a large number of other young stallions, and they train together in the three basic disciplines of dressage, jumping and cross-country. Fifty per cent of the total marks awarded at the end of the whole test are given for the training period, and the 100 days culminate in a competition in which they are assessed for the remaining fifty per cent. Thus to my mind a graded and performance-tested Warmblood stallion is certainly worthy of his mares!

Other breeds are becoming stricter, and this can only enhance the quality of stallions at stud and their future progeny. In England, performance testing, or a proven performance record, is a requirement of some breed societies, though the only official 'approval' of the Thoroughbred stallion is carried out by the National Light Horse Breeding Society (HIS – see p14). It should also be noted that Wetherbys have a complete register of all Thoroughbred, and some non-Thoroughbred, stallions.

It is hoped that the stallion you choose will improve the points of your mare that are weak, and that the bloodlines are strong so that he will 'stamp' his stock. It is known that a mare normally gives a minimum of 50 per cent of her own genes to her offspring, and it could be as much as 80 per cent; thus the mare line is most important in *all* pedigrees for that reason. Stallion and mare must complement each other. The breeder should keep in mind that the mare is capable of limiting the quality of her foal, regardless of the quality of its sire.

Points to look for in a stallion's pedigree

1 Pick out the stallion's sire and results of his progeny.
2 Pick out the dam's line, taking careful note of her dam's sire and his progeny.
3 Study the dam's own offspring and their performance results.
4 Think about the stallion himself and his offsprings' winnings.

This picture was used to support the information given on the stud card of Advanced event stallion, Welton Apollo

5 Pick out the third and fourth generations and consider these very carefully. Their attributes will more than likely be present in the next generation, and will be revealed in either performance, ability, conformation and/or colour. The ancestors of the third or fourth generation will no doubt be apparent in many other pedigrees, and their results should be studied.

Consider all these factors along with your own mare's ability and type.

Line breeding is where the same horse is featured on both sides of a pedigree. It is usually used to increase the impact of a specific horse's genes: for example, in Assert's pedigree the stallions Nearco and Native Dancer have both been used in the fourth generation in both the sire's and the dam's pedigree, no doubt in the hope of adding the fastest and best blood lines of the time. The result was a racehorse whose time, form and winnings were top-class.

It should be noted that it is not advisable to carry out line breeding closer than the third generation, as the result would be deemed as inbreeding.

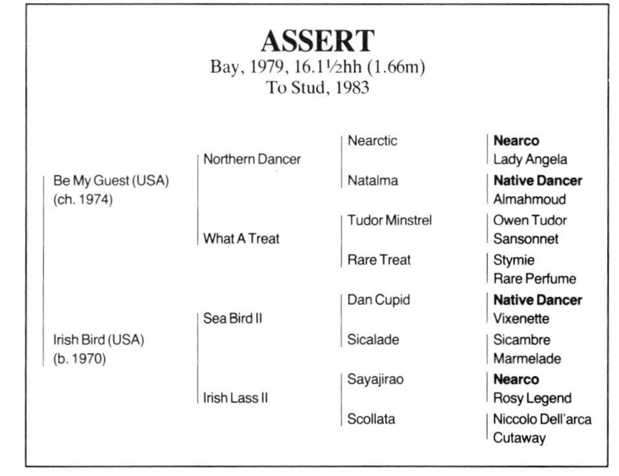

ASSERT
Bay, 1979, 16.1½hh (1.66m)
To Stud, 1983

Be My Guest (USA) (ch. 1974)	Northern Dancer	Nearctic	**Nearco**
			Lady Angela
		Natalma	**Native Dancer**
			Almahmoud
	What A Treat	Tudor Minstrel	Owen Tudor
			Sansonnet
		Rare Treat	Stymie
			Rare Perfume
Irish Bird (USA) (b. 1970)	Sea Bird II	Dan Cupid	**Native Dancer**
			Vixenette
		Sicalade	Sicambre
			Marmelade
	Irish Lass II	Sayajirao	**Nearco**
			Rosy Legend
		Scollata	Niccolo Dell'arca
			Cutaway

A quality Welsh Section A pony stallion stood up for inspection. A stud card should inform you of the stallion's winnings as well as those of his offspring

Those owners who don't have a record of their mare's breeding should remember that their choice of stallion is of paramount importance, as his pedigree is the only known factor other than the mare's conformation, looks and performance ability.

STUD PROCEDURE

The cost of taking your mare to stud

The stud fee for a stallion varies enormously. In the UK, horse stallions (excluding the racehorse) stand from £175 up to £1,000 (often plus VAT), and pony stallions from £50 to £200.

The actual fee reflects several factors, some of which are obvious but are nevertheless listed here:

1 Value of the stallion
2 Quality, performance and paces
3 Pedigree and bloodlines
4 The stallion's winnings
5 Value of offspring, and their winnings
6 Today's market price
7 The area and the facilities offered by the stud.

The average fee is hard to quote, especially for the Thoroughbred – it should be noted that in the UK, the NLHBS awards premiums throughout the country, and those stallions have to stand for a fee between £100 and £200. Each different breed has its own market value – for example in the UK, most approved Warmblood stallions stand for over £350 – though in some cases the stallion owner may give concessions to well-bred or proven performance mares; that means you may be allowed to use the stallion for less than the quoted stud fee.

Stud fees and terms will vary. Listed here are the most usual:

NFFR = No Foal Free Return 1 October: the stud fee will have to be paid on arrival at stud or before the mare leaves (depending on the conditions of the stud), but if the mare is certified not in foal on 1 October you are entitled to take her back to stud the following season when only livery charges will apply.
Straight Fee = Payable on arrival at stud or before the mare leaves; non-refundable.
50% on arrival/50% 1 October Terms = 50 per cent of the stud fees are payable on arrival, and 50 per cent on 1 October if the mare is certified in foal at that time.
NFNF = No Foal No Fee; an unusual agreement, rarely used by studs today.
Livery charges should be listed and, once again, they will reflect not only the work involved but also the quality of the stud's facilities and staff competence/experience. Most studs will offer the following options:

- Mare at grass;
- Mare with foal at foot at grass;
- Additional cost for feeding mare at grass if required;
- Mare stabled at night;
- Mare with foal at foot stabled at night;
- Foaling down fee.

Groom's fee: It is usual for studs to charge a groom's fee, which should be divided between the grooms at the end of the season. (In the olden days, grooms were customarily tipped by mare owners.)

Most studs do not charge for their time if your mare needs to be examined by the vet for gynaecological reasons, but they may add a handling charge when the blacksmith visits or if a veterinary visit is needed because your mare is sick. Similarly, a stud will have to charge for its time if your mare is wounded and needs daily attention. Most studs advise customers if a mare is sick or wounded, **but during your visit to the stud these points should be discussed so that you are fully aware of the financial implications**. Fees for AI may be different, so check before sending off for your nomination (see p39).

Stud visits/Viewing of potential stallions

Having collected and studied the stud cards, it is time to view the sires you have picked out. Several studs today stand more than one stallion and, providing those stallions have been well chosen and are relevant to your mare it will save you some time and mileage to visit this type of establishment. In addition, thought should be given to the facilities and services offered by the stud. Your mare is valuable to you and the way in which she is looked after during her period at stud is important. First impressions of a yard are significant. You will only have a short time to spend with the owner or stud manager and in this period you will have to decide if you are going to be satisfied with the service offered.

SUMMARY: COSTS TO BE CONSIDERED WHEN TAKING A MARE TO STUD

Expected costs at stud:

Stud Fee
Livery charges:
  2–5 weeks at grass (or stabled)
Veterinary fees:
  Two swabs
  Probable EVA blood test
  Possible examination whilst at stud
  Pregnancy diagnosis
Blacksmith, trims
Worming
Groom's fee

Note: The stud will have to take into account wear and tear, overheads, staff or other sickness costs.

Expected costs at home:

Keep at home during pregnancy (see p51)
Foaling down at home with no problems
Foal costs to weaning, including:
  Registration papers
  Worming
  Blacksmith
  Pre-vac T vaccination
  Incidentals
Advertisement for sale

A stud in Kentucky, USA. When you visit a stud with a view to leaving your mare, first impressions are most important

The stud itself

Although the suitability of the stallion that you eventually choose is important, so also is the stud where he stands, *ie* the quality of the premises and the standard of management. There are many locations purporting to be professional, but are they really? This is up to you to discover during your visit. Studs that have been established for some time, have experienced staff, that keep up-to-date with modern breeding trends, and have facilities that indicate the best stable management, may well charge higher livery fees, but justifiably. Large sums of money will have been spent on examination stocks, artificial insemination facilities, post-and-rail paddocks, closed-circuit television and other equipment probably not visible when you visit. **Stud hygiene** is, of course, of the utmost importance in order not to spread infection.

Make an appointment to view, and do be polite and cancel it if you are unable to attend after all. On arrival at the stud you will probably find that the stallion is tied up ready for you to look at. He will be brought out of his stable and stood up for inspection. He will then be walked and trotted in-hand so you can assess the correct-

SUMMARY: POINTS TO LOOK OUT FOR WHEN YOU VISIT A STUD

1. Knowledgeable, helpful staff
2. Good grazing and well-maintained pasture land
3. Post-and-rail (or alternative safe) fencing
4. Stallions and other stock in good condition
5. Examination stocks
6. Visiting specialist stud vet
7. Evidence of good stud hygiene
8. A good customer service.

(Above) *A stallion running free shows his true paces. It is especially important to look at movement if you are considering a young sire who has no progeny on the ground*

(Right) *Good safe fencing is essential for well managed studs*

ness of his movement and paces. Some studs will ask you if you would like to see him lunged, loose schooled and jumped or ridden (if applicable). Note carefully all the points about the stallion that were listed earlier in this chapter; most important, does he have presence, a good temperament and a kind eye?

Remember to take a photograph of your mare with you to show the stud manager – this may assist when you are looking for the traits that you need to better your mare's offspring. If you have never seen any of the stallion's youngstock, request to do so now, and if none are available, the manager should be able to give you an idea of what type of foal you could expect from this particular stallion. He may have some photographs, or you may be able to inspect them at other yards.

EXAMPLE NOMINATION AGREEMENT

I ....

agree to take a nomination to the stallion ....

for the season 199_@ £ ....(plus VAT) Terms (see P 2)..

Mare's name .... Breed .... No. ....

Age .... Height .... Colour ....

Sire .... Breed .... No. ....

Dam ....Breed .... No. ....

Foal at foot **YES/NO** Date born.................................. Sire ....

Breeding history 199_ ....

Does the mare normally conceal her heat? **YES/NO** Has she had a Caslick? **YES/NO**

Has the mare had any infectious or contagious diseases? **YES/NO** ....

Has the mare any known permanent lameness or hereditary condition? **YES/NO** **If yes please give details.**

....

Date of mare's last vaccinations: Influenza .... Tetanus ....

Date of clitoral swab (C.E.M.) .... Date of cervical (endometrial) swab ....

Any other information that may assist the Stud ....

....

Registered owners name ....

Name and address for accounts and correspondence ....

....

Tel. no. (day) .... Tel. no. (eves) ....

Address where mare normally kept ....

.... Tel. no. ....

Veterinary surgeon's name .... Tel. no. ....

Page 1

Page 2

TERMS AND CONDITIONS

N.F.F.R. 1st October means No Foal Free Return - if mare certified not in foal by 1st October she will be eligible for a free return during the following season.
Straight fee means if mare does not conceive whilst at stud, no payment is re-imbursed.
A.I. means the provision of semen by artificial insemination. There is an extra charge for the collection of semen. 20% refund if mare certified not in foal 1st October. Deposit on Equitainer_______.

Veterinary Fees
All veterinary charges for visits, examinations and scans etc. are to be made payable to_________ and handed to_________Stud ten days prior to collecting the mare, unless paid in cash.

Blacksmith
All mares will have their feet trimmed on a regular basis and the owner will be charged accordingly. All mares should have their hind shoes removed before arrival at stud. Please leave front shoes on.

Worming
Mares and foals will be wormed on arrival at stud and every four weeks thereafter. The owner will be charged accordingly.

Swabs
Swabs can be taken at home or whilst the mare is at stud, but no mare will be covered without

1) a certificate to show that she has been examined for Contagious Equine Metritis organism and the results are negative. (Please note results can take up to 10 days.)
2) a cervical (endometrial) swab taken when in season. This swab can be done when the mare is at stud.

Vaccinations
All mares must be currently vaccinated for equine flu and tetanus. Vaccination certificates must be shown on the arrival of the mare.

MARE OWNER'S DECLARATION
I hereby declare that I have read the terms and conditions above and accept:-

1) that _________Stud will endeavour to take all safety precautions with my mare .... and I agree to leave her in their care at my own risk.

2) to give consent for_________ Stud to call for veterinary attention , at my cost for my mare .... without first informing myself, if they feel that it is the necessary course of action.

3) that the stud fee will be paid on the arrival of my mare .... at stud.

4) that the livery charges and veterinary fees will be paid in full ten days prior to the mare .... departing from stud, or in cash at the time of departure.

Signed .... Date ....
(Mare owner's signature)

Once you have viewed a particular stallion (noting at the same time the stud's facilities) you will have to indicate if you have any interest in him as a sire. If you do think he will be suitable, the stud manager will discuss the following:

1 The suitability of the mare in relation to the specific stallion by way of type, conformation, size and breed

2 The stud's administration procedure

3 The relevant fees, including keep charges

4 The stud's veterinary policy and price list

5 The nomination form (a declaration form stating that you wish to use a certain stallion's services, and that you agree to abide by the stud's terms and conditions). Some stallions are only allowed a few mares each season (maybe due to competition commitments) and in these cases, as with very famous sires, you need to reserve a 'nomination' early in the season

6 The stud's policies vis-à-vis length of stay and scanning procedure

7 Covering certificates/foal registration.

The stud will also need to know that your mare's vulva and genital tract are free from infection; this necessitates a **swab**, taken by a vet (either your own, or by the stud vet) – this is discussed in more detail on p22.

By this time you should have all relevant information and can compile your findings, making the final choice well before the actual time that you need to take your mare to stud.

SUMMARY: BREEDING CRITERIA

1 The mare is of a correct age for breeding
2 The mare's conformation and pedigree have been considered
3 The chosen stallion is suitable in breed, type and size
4 The foal will be eligible for papers
5 The stud has good facilities and knowledgeable staff; you would feel happy to leave your mare in their care
6 You have considered the future and have made appropriate arrangement for the mare's gestation period and subsequent foaling
7 You have considered the over-all financial commitment (stud fees, veterinary charges, livery charges and transport costs).

Covering certificates/Foal registration

One more important question to ask is if you will receive a covering certificate, which can later be used to obtain registration papers for the foal. In previous years many British breeders have not concerned themselves with registration papers. There are thousands of horses and ponies in the UK with no form of identity other than that completed by the vet for flu and tetanus vaccination purposes; their breeding is unknown. Elsewhere in Europe, however, registration and branding has been enforced for some years, and in England today registration papers are becoming of paramount importance. No foal born in 1993 or after may compete in affiliated competitions unless it has a registration document. Make sure, therefore, that whatever stallion you choose is registered himself, and that the stud will be able to issue the relevant paperwork to secure you the foal's own paper. **Alternatively** if your mare is registered with a breed society you will be able to acquire papers through that particular body. Whatever the situation, be quite clear in your own mind that the covering certificate that the stud will issue will enable you to register your foal (see p112). **Note** that normally covering certificates are issued at the end of the season and before 1st October. However, if you need the certificate earlier (sometimes required in the show ring) a request should be made to the stud, though the certificate will only be issued if all payments have been made.

THE BREEDING CYCLE

You have visited the studs. You have chosen the stallion. What else needs to be attended to?

The mare's seasons

A mare can only be covered when she is in season (she is then described as being in oestrus); this is the period when she will 'stand' for a stallion (ie submit, usually willingly, to be covered). She will probably show her first season when she is approximately 18 months old; thereafter she should have a season every twenty-one days until September or October. Mares have seasons when the weather is warmer and light days longer, each lasting about five days (though they can be as short as three to four days). During the colder winter months a mare does not normally come into season (she is then 'in anoestrus').

In some parts of Europe, for breeding purposes a stud will try to extend the period of season by standing the mare in artificial light and under heat lamps – if she has had a late foal, she can then be put back into foal the same year, rather than wait for the following spring. Also, some Thoroughbred yards will induce a season by doing this. Before your mare goes to the stallion, **keep a record of the dates of her seasons** – this will assist the stud.

If you have a mare that seems to be in season all the time, have your vet check her. Likewise, if you are considering breeding from your mare and you have never noticed her in season, ask your vet to check her internally and make sure all is in order. The fact that you have never noticed when she is in season could be simply because she has what is known as a 'silent season', where she ovulates normally but shows no external, behavioural symptoms of doing so. This is not unusual, and is highly likely to change when she goes to stud and is shown in person to the stallion, who will stimulate her behavioural reaction.

The foaling heat

This is a term used to indicate the mare's first season after foaling down; it normally occurs 7-10 days after the foal is born and this particular season lasts only 2 to 3 days – unless a stallion is at hand it is often not noticeable. Thereafter the mare's system will return to its usual season every 21 days.

Details of a mare's seasons (the oestrus cycle)

Each cycle should last 21 days: 5 or 6 days in season (oestrus) and 15 or 16 days out of season (described as dioestrus). Today, stud vets do tend to discuss hormone changes during the oestrus cycle; this table may assist in understanding their terminology.

			Hormones Dominant	External Signs
Day 0–6	In season	Oestrus	FSH and OES	Tail, up, winking, relaxed cervix
Day 5	In season	Ovulation	LH	Tail up, winking, splayed hind legs, urinating, open cervix
Day 7*	Out of Season	Dioestrus	PROG (and FSH at day 13)	Ears back, kicking, closed cervix
Day 20	Start of next season	New oestrus	FSH and PROST	Renewed interest in stallion

*If the period of oestrus lasts longer than 8 days this indicates that there may be something wrong.

FSH = follicle stimulating hormone
PROG = progesterone
PROST = prostaglandin
LH = luteinising hormone
OES =oestrogen

Prostaglandin injections

This treatment is commonly known as **a PG**. A mare can be brought in to season if the vet administers prostaglandin (which is a substance that the mare produces naturally when she comes into season). If your mare goes to stud and in relation to her predicted oestrus, shows no sign of coming into season after approximately ten days, the stud manager may well recommend the use of prostaglandin. The injection should bring a mare into season in about two to four days (or sometimes longer after administration). However, it is not always effective (due to the fact that it should be given five days after ovulation) but a second injection can be administered seven days later.

It should be noted that this injection sometimes causes a mare to sweat profusely, and in rare cases can cause mild colic. Naturally a stud will keep a mare that has been given a PG under observation for a short period.

Regumate is used to control the oestrus cycle of cycling mares, but at the discretion of the vet; it is an expensive but effective remedy in cases when mares do not have regular seasons. It is a liquid substance that is added to the daily feed for a 10-day period, and it *must* be used as directed by the vet in order to have the required effect. The majority of mares will return into season naturally within 8 days of the last oral administration,

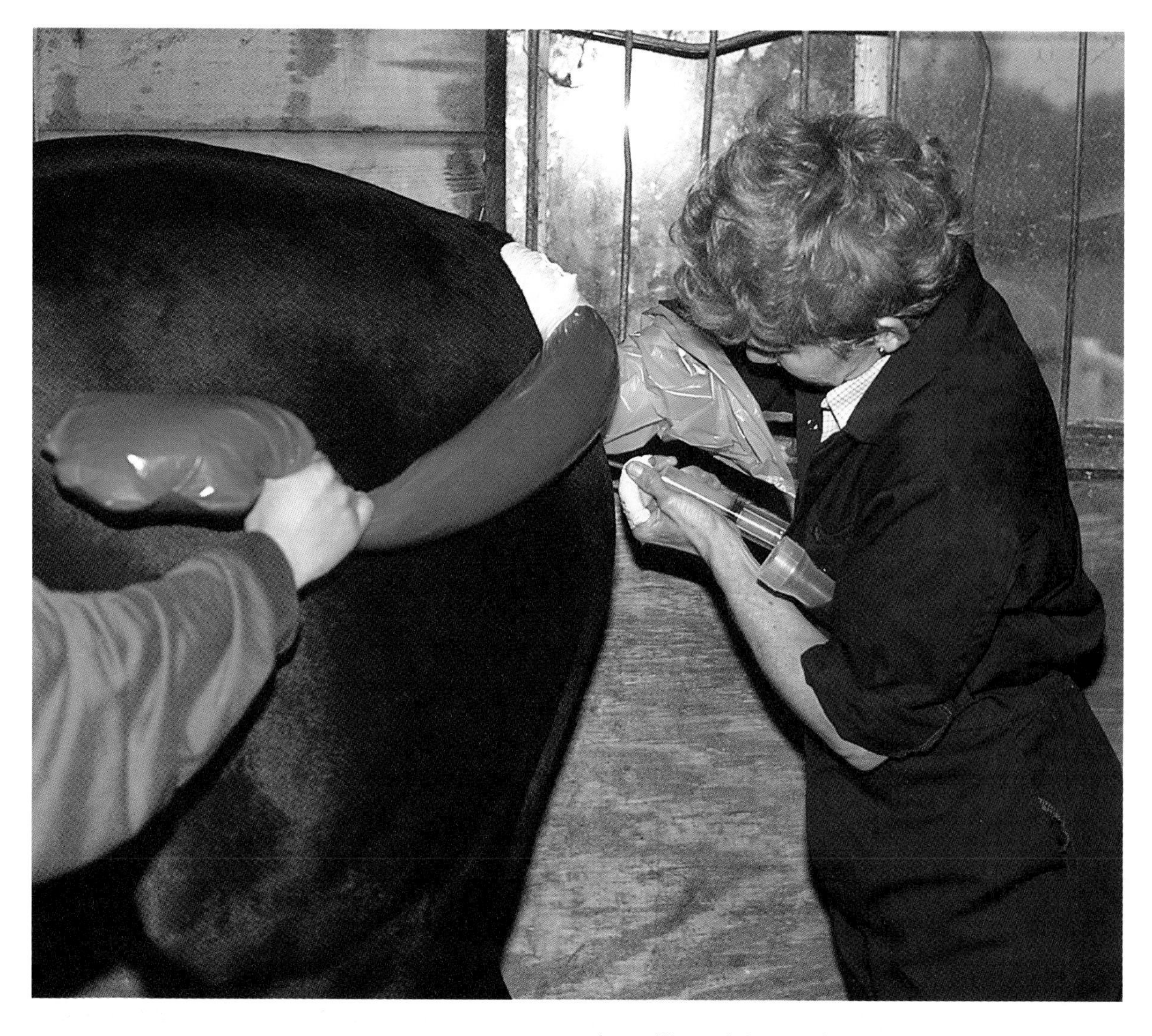

A mare being swabbed in her own stable before going to stud

but in some cases a prostaglandin injection is given on the 11th day and the season should start within 3–4 days.

Your mare's swabs

Today there are few studs that will allow their stallions to cover mares that have not been swabbed. Swabs can be taken at home or at the stud, and there are two different types:

The clitoral swab (commonly known as the CEM swab). This may be taken at any time (*ie* the mare does not need to be in season) and the result will take 7–10 days. Your veterinary practice will send the swab away to be cultured and it actually has to be in the incubator for several days; hence the reason it takes time before a result can be known. If you have the swab done at home, remember to get a certificate to take to the stud. If you send your mare to stud without having had the swab taken, she will be there for at least ten days before they can consider covering, even though she may come into season during this period.

It is rare today to find a mare where the result is positive. However, continued swabbing of mares will ensure that the disease known as **contagious equine metritis** (CEM) is kept out of the UK.

The cervical swab (also known as 'uterine' or 'endometrial'). This has to be taken **when the mare is in season**. If the result is positive, which is quite common, the mare will need treatment. It is normal for this particular swab to be taken when the mare is already at stud. Reasons:

a) it is without doubt easier for the stud to know when she first comes in to season so, with the mare being at stud, the swab can be taken straightaway;

b) there will probably be a vet close at hand to do this;

c) the result only takes 24 to 48 hours; therefore if all is well and the swab is clean, then the mare can be covered during that particular season;

d) if treatment is required, it can be done at the stud;

e) on occasions, if the result shows a mild infection unique to the mare, the stud vet may advise covering of the mare, and give treatment after covering, known as a **'post service treatment'**.

The cervical swab relates to the gynaecological condition of your mare. If the result is positive, and this can be so even in maiden mares, then she is less likely to conceive without treatment. Unlike the CEM (clitoral swab), this condition is unlikely to harm the stallion if he covers her, although it is recommended that he should not cover a different mare for at least 24 to 48 hours after.

SUMMARY:
SWABS – TWO TYPES

1. **The clitoral swab** (commonly known as the CEM). Takes 7–10 days for a result. Can be taken when the mare is not in season and before foaling down.
2. **The cervical swab** (commonly known as the 'uterine' or 'endometrial' swab). **Must** be taken when the mare is in season. Unless taken by a non-stud vet, results will be available within 24 hours.

There are varying opinions as to whether both swabs are necessary in all cases – you may find some studs which make exceptions to the rule. These exceptions may include the following:

Maiden mares: If bred at home and guaranteed *never* to have been covered, then the clitoral swab may not be required. Also, as a maiden mare is a virgin and the cervical swab requires penetration of the hymen for the first time, this one may not be required either.

Mares with foal at foot: If the mare has a healthy, live foal at foot then it is unlikely that she has contracted equine metritis, and the clitoral swab may not be required.

Note: The majority of well run studs require both swabs to be taken whatever the circumstances; please respect their wishes – in the end it should prevent any outbreaks of venereal diseases in this country.

The stud vet

Most studs today have access to a specialist stud veterinary surgeon who will visit on a regular basis; the large Thoroughbred stud will have a live-in stud vet who will be on hand for difficult foalings and daily stud practice. Check the availability of a stud vet during your visit.

Choosing the right time of year to take your mare to stud

Circumstances often dictate this, and the subject is usually discussed when you visit the stud.

Maiden mares: The timing in this case is your decision, though nature shows us that the best time for a foal to be born is in the spring. Reasons: the climate is warmer and the grass is greener, and **spring grass** stimulates the production of a good milk supply due to its high level of protein. Some people take their mare to stud as early as March or April (or as soon as the stud can accommodate her), and Thoroughbreds required for racing will go even earlier. However, bear in mind that in the more northerly regions of Europe the spring begins much later, so the month of May is perhaps more appropriate. If you send your mare to stud in May and she takes (conceives) during her first season at stud, the foal will be born in April – an ideal time of year when he will have every chance to grow strong, and when the mare will benefit from the best grass of the year. And if you want show your mare and foal, then he will be big and strong by the time the show season begins.

Disadvantages of an early foal: It is imperative that you have the time and facilities to keep a good eye on your mare and foal – March in more temperate climates can have extremely cold, wet days, and mare and foal will almost certainly have to be stabled at night, or at least provided with a shelter. These facts could well deter you from taking your mare to stud until May. If so, fair enough, but none the less remember two things: the mare may not conceive straightaway, so you could end up with a later foal than you had intended; and if you want to show mare and foal you will not want a late birth.

Disadvantages of a late foal (*ie* born after 1 July):

1 No spring grass (the ingredient so vital for the mare's production of milk) so you will have to supplement her diet with hard feed
2 The weather could be very hot with many flies – not pleasant for a young foal, and a medium for spreading disease
3 Summer colds are often rife, and a young foal can be extremely susceptible to these.

Mares can conceive well in to August, although the success rate is lower and stallions are known to be more fertile in the spring.

Do you wait until your mare is in season, or just take her to stud when it seems convenient? Unfortunately, travel and/or the combination of a new 'location' often has an effect on the mare's cycle, so if you decide to wait and take your mare to stud when you see her come in to season, it can happen that when the mare is shown to the stallion the next day she is found to be no longer in season. What a blow, but such is nature. However, if you are on a very strict budget it is certainly worth a try, as the mare need only stay for the period that she is in season (see p29).

The alternative way, which I advise, is to calculate when your mare is next due in season and send her to stud just a few days before. In this way she will become accustomed to her new environment and can be introduced gradually to her future 'husband' before the time is right for covering. You will find that most studs would welcome this – it makes their job easier because they can then get to know your mare, and how she reacts to the stallion, both before and as she comes into season.

Mares in foal: If your mare is already in foal and you have planned to foal her down at home you will only be able to make provisional arrangements with the stud as to when you will take her – first you will have to know that all is well with the birth, and that the foal is strong and healthy. Once she has foaled, you will have to decide if you want her covered on her foaling heat, or on the next one, three weeks later. I advise to wait for the latter, although there are some mares who conceive more easily if covered during the foaling heat. The current trend is to cover on the later season (*ie* not during the foaling heat); the mare could have been torn, and particularly if she had a traumatic foaling it is less stressful for her if you wait a little longer before covering again – **consult your vet** on this issue. Also the mare is often *very* foal-proud at ten days, therefore it can make covering more difficult for the stud.

Remember that if your foal has been born quite late into the summer, your mare will be approaching the state of 'anoestrus' *ie* she will be less likely to come into season, or ovulate, at all. In this case the stud vet may well advise to use a PG (prostaglandin) injection, which should bring her into season (see p21). So you would take your mare to stud when the foal was approximately two weeks old, and the mare's season would be induced artificially (at two weeks the foal should be better able to adapt to a change).

The mare's condition when she goes to stud

Your maiden mare (or mare without foal at foot) should be in a good physical condition when you take her to stud. It has been established that the body condition of mares at the time they are bred from can affect conception. It is imperative that the mare's bodyweight and condition is maintained – it is known that by doing so, the rate of conception is higher: mares that are underweight, or very overweight, are likely to have less chance of conception. If your mare has been competition fit, prepare to rough her off gradually; psychologically and physically she should become a brood mare, but *do not* allow her to lose condition.

Most visiting mares are put out to grass during the summer months, providing the stud has good grass, or fed additional protein if required. You may wish your mare to be stabled at night, but it is likely that she will be put out to graze during the day. However, try not to take too many rugs – if the time is early spring, one night rug and a New Zealand are acceptable.

Taking a fat, unfit mare to stud is also not advisable – if she is very overweight she may be more difficult to get in foal.

SUMMARY: THINGS TO REMEMBER PRIOR TO TAKING YOUR MARE TO STUD

1. Visit stallions and studs that you consider a possibility.
2. Make a decision as to which stallion you are going to use.
3. Is the location acceptable?
4. Advise the relevant stud manager/owner and reserve a nomination.
5. Arrange for one or both swabs to be taken.
6. Rough your mare off if applicable, *but* maintain her condition.
7. Check that your vaccination certificate is up-to-date – the stud will probably wish to look at it.
8. Check with the stud about worming. Most studs worm every mare on arrival and regularly thereafter.
9. Decide when you are going to take your mare to stud.
10. Keep a record of your mare's seasons.
11. Decide if you are going to leave her at stud until she is confirmed in foal, or bring her home after covering.
12. Decide what you are going to do with her when she comes home *ie* is she going to become a brood mare, or are you going to continue to ride her? (see p47).

2

THE MARE AT STUD

Ten years ago taking a mare to stud was a much simpler exercise than it is today, but the success of conception was far lower. With the advancement of research and the availability of the **specialist stud vet**, fewer mares leave the stud without conceiving. So how much stud 'jargon' do you really need to know? Outside the breeding circle you certainly would not come across the words 'teasing', 'trying wall', 'AV', 'prostaglandin', Regumate and other brand names together with specialist terminology. And what is involved if the stud does have difficulty in getting your mare into foal – will it cost you an awful lot more? What should you remember to take with you when your mare goes to stud, and what actually takes place on a day-to-day basis?

If you have a greater understanding of these terms, and also the practical implications of the more specialist stud activities such as artificial insemination and embryo transfer, you will be better able to talk matters over with stud personnel. This chapter examines how long the mare might be expected to remain at stud, and takes a brief look at the 'walk-in stallion station' method; it concludes with an insight into the various methods of pregnancy diagnosis, the advent of the vet's scanner, plus a short piece on twinning and pregnancy failure.

THE STUD AT WORK

What happens during the time your mare is at stud?

First and foremost, let me say – as an owner of a stud standing a number of stallions – that there are two priorities vis-à-vis visiting mares: 1) to look after your mare to the best of our ability during the time she is at stud, and 2) to endeavour to send her home in foal.

THINGS TO REMEMBER WHEN YOU TAKE YOUR MARE TO STUD

1 The completed nomination form (if not already sent in advance).
2 Vaccination certificates.
3 Relevant swab certificates.
4 The mare should be in good condition physically (see p24).
5 Remove the hind shoes (removing the fronts may not be necessary, but check with the stud).
6 Send the mare with a well-fitting comfortable headcollar that will not rub. She may have to to wear a headcollar all the time whilst out at grass (a means of identification for the stud).
7 Advise the stud if your mare is unsound, has any recent injury (see p24), or is on medication.
8 Advise the stud if your mare has any known allergies to feed, drugs or wormers. (There may well be a section on the nomination form which will relate to this point.)
9 Advise the stud of your mare's previous breeding history (if applicable), and of any specific problems that have been encountered in the past.
10 Dates when your mare was last seen in season (if applicable).

It should be noted that whilst your mare is at stud **you must leave management to the stud to which you have chosen to send her**. In this section I have tried to paint a picture of the stud at work so as to give you an idea of what takes place, from the time that you take your mare to the stud until she returns home. I hope it will help you to better understand the terminology involved, and also the problems that a stud might have in getting your mare in foal.

On arrival: When you arrive with your mare at stud, someone will be allocated to receive her, and she will probably have an identification tag put on her headcollar. You will then be asked to produce the necessary documentation required by the stud (as listed above). Make sure that you ask if it is the stud's policy to telephone you about your mare's progress, or whether you should check with them. Each and every stud varies in its administration procedures – it is up to you to find out what they are.

The period at stud can be divided into three parts: the period prior to covering; covering; and the period after (and in between) covering.

The period prior to covering

The stud's main aim during the period prior to covering is to get to know your mare, her mannerisms, and her reaction to the stallion. During the first few days at stud your mare will gradually become accustomed to her new environment; it is likely to be quite a mental change for a maiden mare to be put out to grass if she has been accustomed to being in work and stabled. However, she should adapt easily and will soon learn that the main excitement of the day is the appearance of the stallion, or being taken to the **trying wall** for teasing. The stud term '**to tease**' means the chatting up of a mare by a stallion. Today most stallions are deemed too valuable to be allowed the freedom to run out with their mares, so a method known as 'teasing' is used to find out when a mare comes into season (into 'oestrus' which comes from the Greek word meaning 'mad desire').

At most studs a **teasing** or **trying** wall is used. It should be a well made, solid wall built in the centre of a fence line or in a specially designated covering paddock and, most important, it must be constructed of a material that will do no harm to the mare, if and when she kicks at it.

A '**teaser**' is the name given to a stallion who is only used for teasing and **not** covering. 'Teasers' are predominantly pony types used in Thoroughbred studs where they do not want their stallions put under any unnecessary stress. They will be well trained to 'chat up' the mares, and most teasers are allowed to cover once or twice before the end of the season.

A mare being tried at the teasing wall and showing classical signs of being in season

The teasing procedure is as follows: the stallion is stood up on one side of the wall, being held in-hand by the handler (who in most studs will be known as the **stallion man**), and the mare is led up to the wall on the opposite side. She will probably be wearing a headcollar, and the handler will use either an extra-long rope or a lunge line so that if she reacts violently towards the stallion he has ample room to keep out of the way. In all studs, the safety of the staff is of paramount importance, and at times, handling of stallions and mares can be dangerous, so safety precautions are put into practice as a daily routine. If the stud is dealing with very big strong mare, it is likely that they will use a bridle.

Once the mare has been brought up to the wall they will be allowed to sniff at each other. A mare who is not in season will most likely take an instant dislike to the stallion, lay her ears back, show her teeth, clamp her tail down hard and/or strike out at the wall. If, on the other hand, she stands quietly, the stallion will be allowed to sniff at her hindquarters (*ie* tease the mare) and after a time she may lift her tail. This is a sign that she is either coming into season or on the way out.

When a mare is fully in season she will normally positively enjoy being next to the stallion and will lift her tail, crouch down slightly and straddle her legs, urinating at the same time: this is the period when it is likely that she will 'stand' for the stallion (allow covering to take place). Accurate detection of when a mare is properly in season is therefore of paramount importance, and every mare is different: each has her own pattern of behaviour, though experienced stallion men or stud personnel are able to recognise different reactions and, in the majority of cases, interpret the signs accurately. There are mares that show in season without having a stallion in the vicinity – they may well react to another mare or gelding in the field or may call to a stallion or gelding, and in some cases, mares in season tend to walk the fence line; but there are others that are much more difficult and do not show easily.

SUMMARY: CLASSIC SIGNS OF MARES IN SEASON (IN OESTRUS)

The mare may show one or several of the following points:

1. Lifting of tail combined with:
2. Swelling of vulva lips
3. Mucus will escape from the vulva
4. Winking (an action that pushes out the clitoris)
5. Crouching down as she straddles her legs and urinates.

Teasing is normally carried out every other day. Some studs walk a stallion past the mares' fields so that they can observe the mares' reaction to the stallion when in a group. For example, a 'shy' mare may show in season in her normal environment but not when taken up to a trying wall, and vice-versa.

There are many anomalies, and it is the stud's task to determine when a mare is well in season and ready for covering.

There are other locations where teasing can take place, but these could be hazardous, depending on the circumstances and location. Teasing is sometimes seen conducted over a stable door – difficult if there are overhanging eaves or if the stallion attempts to crawl over the door in his excitement at being presented with a mare in front of his stable.

Individual mares react differently. If your mare was expected to show in season within a few days, but has not done so having been at the stud for a week or more, it is highly likely that the stud vet will suggest that he gives her a prostaglandin injection (PG) a hormone injection to hasten the oestrus cycle and which should bring the mare into season within the next three or four days.

The stud vet

The stud veterinary surgeon (generally known as a stud vet) will have studied the reproductive systems of both mare and stallion in far more detail than the general horse vet. As we have said, scientific research into stud work advances steadily; recent results have meant that the vet can manipulate the mare's seasons, he can now treat uterine infections successfully, and can assist the problem mare to carry a foal full-term. He should have up-to-date information on the latest scientific research, in addition to his knowledge of new drugs, treatments and the latest thinking on the gynaecology of the mare.

As soon as the mare does show in season a **cervical** (also known as 'endometrial' or 'uterine') **swab will be taken** (see p22); this can only be done when the cervix is open during oestrus. At many studs, mares are placed in **stocks** for examination – examining mares and taking swabs is made much easier for the vet today with the availability of **stocks**. These are similar to a cattle crush in construction, and are used to restrict a mare's movement; they have two solid sides or bars, with a door at the front and back. To encourage the mare to walk in, the front door is left open – as she enters it is quietly closed, followed by the closing of the other door. The stocks should be well finished and padded for safety, with the strongest of fixing into the ground.

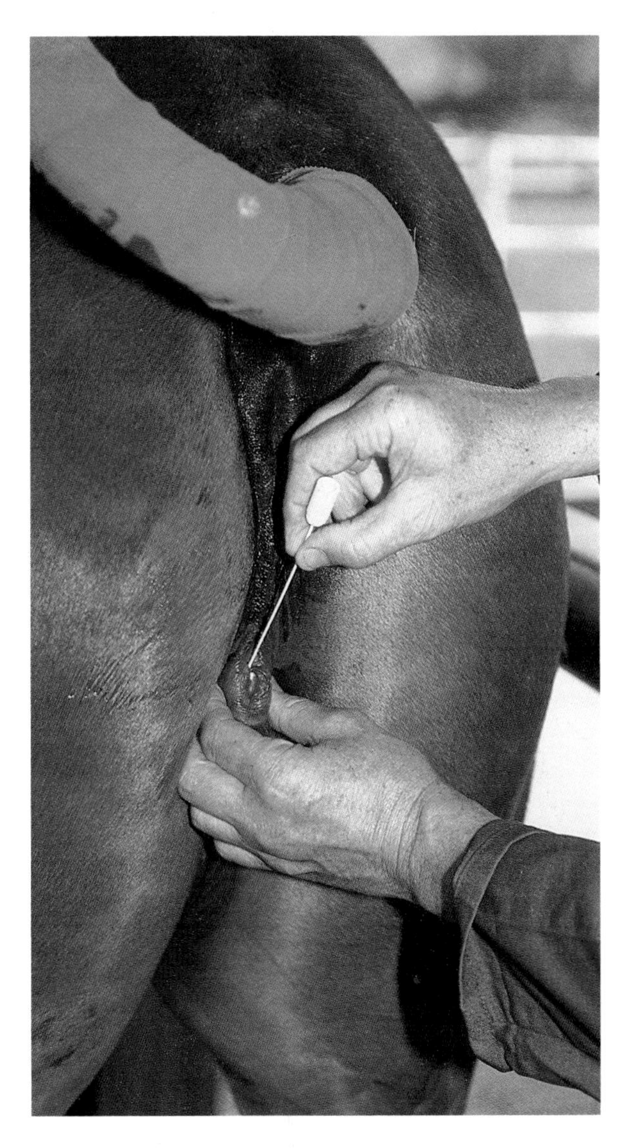

The vet taking a cervical swab

Once the mare is in the stocks she can be fed some hay or given a feed whilst the stud vet carries out his duties. If no stocks are available the stud will probably need two or three handlers to restrain the mare whilst the swab is taken. A maiden, in particular, may well be worried about the procedure, especially if a cervical swab is required, which necessitates the breaking of the hymen. A twitch is recommended in these circumstances, unless she reacts violently to it, in which case the only alternative is to use a mild tranquilliser.

(Left) *A mare standing in stocks awaiting examination by the vet*

Provided the cervical swab results are clear (and these will be obtained within 24–48 hours) the stud will be ready to cover your mare. If the result is positive the vet will suggest a treatment which may be given before or after covering (commonly known as pre- or post-covering infusions), depending on the findings. There are some results which show that covering should not take place at all until a series of treatments has been given. In such a case the stud will probably advise you of the facts, and wait until the next season for covering as long as the infection has cleared.

Scanning

The ultra-sound scanner is used widely today, not only to determine states of pregnancy, but also to show the state of the ovaries, and the uterine oestrus pattern, from which a stud vet can estimate when the mare is likely to ovulate; in other words the use of the scanner enables him to know when is the optimum time to cover. In the large Thoroughbred stud where a stallion may have to cover up to four mares per day the scanner is particularly useful, since he will then only cover a mare at the time designated by the vet. The time of covering is crucial, as conception only takes place **when the follicle is ready to burst**.

The scanner assists enormously with the difficult mare who may either not show well to a stallion, or stays in season for an extended length of time. The scanner is now well established as the stud vet's most useful tool.

However, using the scanner is a costly exercise; in the smaller stud its use is normally restricted in order to keep costs down for the mare owner, and it is only employed in difficult cases.

Covering the mare

Having decided that the mare is ready to be covered she will be prepared by:

1 Putting on a headcollar or bridle;
2 The vulva and surrounding area will be washed clean with warm water;
3 Applying a clean tail bandage or tail guard;
4 Kicking boots will be put on her hind feet so

A mare fully in season and ready for covering

that if she does lash out at the stallion, the impact will be softened by the boots;

5 A twitch will be placed ready for use if considered necessary.

The mare is now ready for the stallion who will be brought back to the wall to tease her again; at this point the stallion handler will decide whether to apply a twitch as a safety precaution – it may assist in calming the mare if she is worried. Hobbles are only occasionally used and certainly by the most capable of personnel; in very difficult cases, a mild tranquilliser might be administered. The majority of coverings are easy, and the mare will enjoy the experience.

Once the mare has been covered, the boots and twitch (if applicable) will be taken off and she will be walked around for a few minutes to prevent her from standing and staling immediately after penetration.

If the mare has a foal at foot, the stud manager will decide whether to leave the foal in the security of the stable during covering, or to hold him nearby (but well out of danger). Studs vary in their management policies and the decision will depend on several circumstances, including the age and manners of the covering stallion, and the mare herself. Some mares are very foal proud and even if fully in season, may fret if their foal is not in sight.

Whilst your mare is in season she will probably be covered every other day from the second or third day of her season. However, I cannot generalise here as so much depends on circumstances. If a stud vet has been called in, he will advise the stud when to cover and your mare may well only be covered on one or two occasions, at the optimum time. It is interesting to note that sperm remains active in the mare's reproductive tract for at least 48 hours and maybe even longer, but in order for fertilisation to take place the mare must ovulate during this period and the time of ovulation is approximately 24 hours before the end of the true season.

A very small percentage of mares will not allow a stallion near them even if they are fully in season. Some years ago such a case was probably sent home, or possibly presented to an alternative stallion in the hope that she might like him better! Today however, with the availability of the scanner to confirm that the mare is in full 'oestrus' and with artificial insemination, there is the possibility of conception. Artificial insemination (commonly known as AI) is explained in full on page 38.

Nature's way of copulation

This entire procedure, from teasing through to actual covering is, of course, carried out in the wild as a natural process, one which I will de-

(Opposite above) *A mare being covered;* (Opposite below) *Washing the stallion off after covering helps prevent spread of infection*

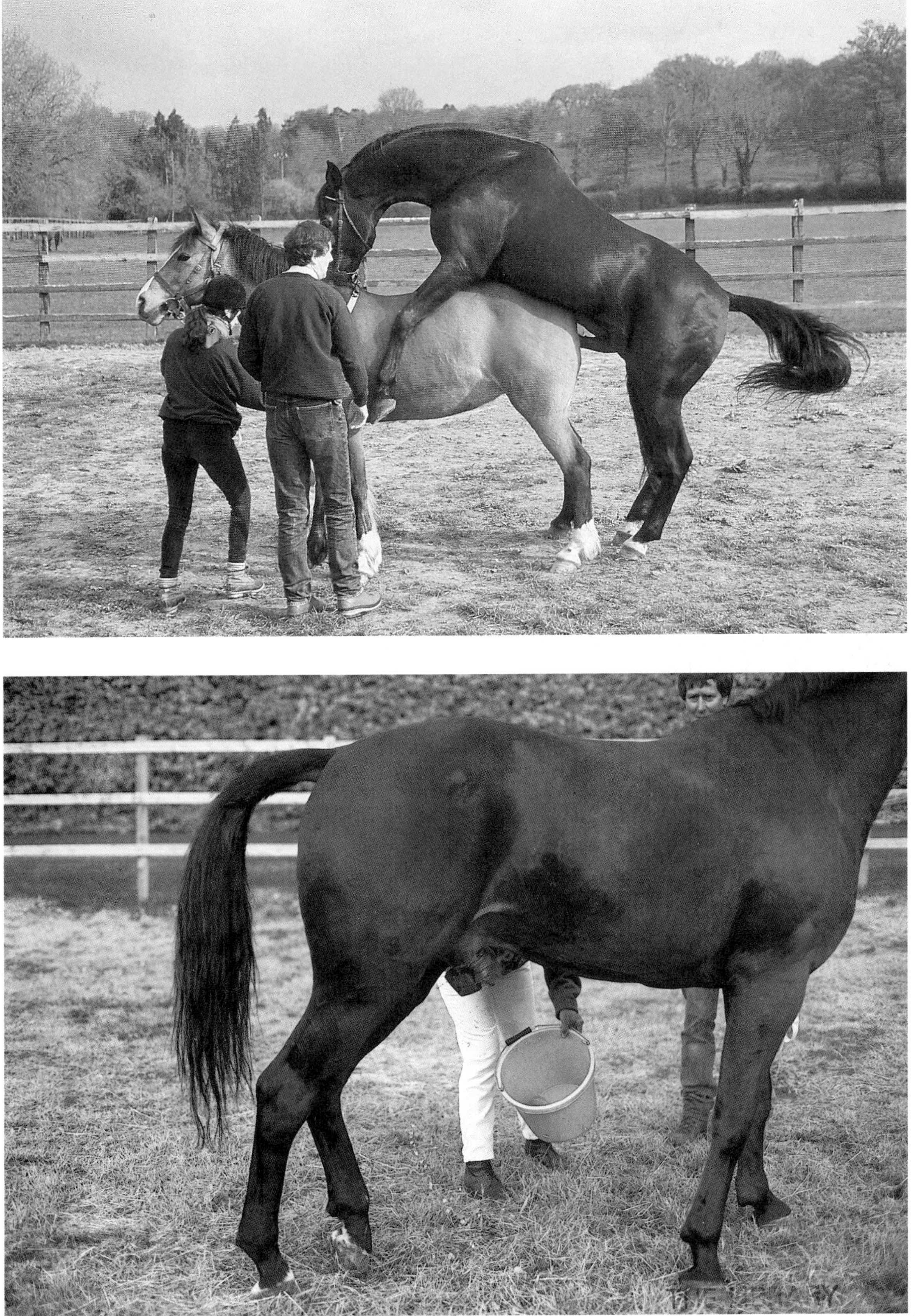

Flehmen posture

scribe here so as to demonstrate that a stud's standard procedure imitates, as closely as possible, nature's way.

Wild horses group together in very small herds comprising the head stallion, several mares and fillies and only very young colts (18 months old or less). Once a colt is deemed a possible sire by the lead stallion he drives it away, and it has to find other companions (as a result, two or three colts are often seen roaming on their own).

Courtship commences when one of the mares begins her season and she becomes receptive to the stallion's advances. He will be heard to call and snort; he will circle the mare, come up close and then bite at her mane and hindquarters. He will almost certainly be seen to curl up his upper lip, showing off his top teeth and sniffing the smell of his lady in season – this position is known as the '**Flehmen posture**'.

As the chatting up continues, and if the stallion is sure that the mare will be receptive, his penis will slide from its sheath and stiffen and he will rear up on to the mare's hindquarters and enter her. Once he has ejaculated his sperm into the mare's uterus he will stay quite still for a few seconds before dismounting. In the wild, stallions may be seen to cover their mares four or five times a day, although scientific facts have proven that this is not necessary for fertilisation.

After covering

Once the mare is no longer in season, she has no more need to be teased by the stallion until she is due to return in season, approximately 16–18 days from the last day that she was seen in oestrus. During this period she will be left in peace to enjoy her natural habitat with just a daily check, although good stud practice will ensure that if a mare is seen to show signs of returning in season prematurely or if mucus is visible from the vulva indicating an infection, then action will be taken and a vet called in for examination.

Although teasing is not practised during this dormant period, if there is a stallion in the vicinity and the mare remains in the stud environment, then it is more likely that any abnormalities will be spotted quickly and acted upon if necessary. If the mare does have a foal at foot it is, of course the stud's duty to ensure that mare and foal are observed on a twice daily basis and nothing left to chance.

It is at this point in time that you, the owner, have to make the decision as to whether you take your mare home after covering or leave her at stud until a positive pregnancy diagnosis is made. It is estimated that the reproductive efficiency on studs during a breeding season averages 80 per cent, and the statistical record maintained by Weatherbys (UK) indicates a live foal return of about 65 per cent amongst Thoroughbred mares at stud (UK figures). The percentage is slightly higher in non-Thoroughbred breeds; this may be because they are not so highly bred (*ie* a TB is bred for its bloodlines, sometimes regardless of a poor physical shape for breeding, and also because they are not expected to breed in February, which many TBs are, when of course normally their oestrus cycle is dormant (in anoestrus).

Walk-in stallion stations

These are very common in parts of Europe, and will no doubt become more popular elsewhere mainly for financial reasons: by using this method livery fees are eliminated, which can represent at least 50 per cent of today's stud fee, if not more. Providing the stud can cater for you on arrival, there is nothing to prevent you from having the required swabs carried out at home, using a stallion or rig that you have on your own yard to check that the mare is in full oestrus, or calling in your own vet to examine her with the scanner. If he gives the go-ahead, all you have to do is load up your mare and if she is receptive on arrival at the stud, she will be covered – if nature does its job well, your mare could go home pregnant.

It sounds so simple – but unfortunately in practice it so often is not.

The most usual problem is also an embarrassing one. You arrive at the stud as arranged, and when your mare is shown to the stallion she will not stand to be covered, for whatever reason. Maybe she was in season at home, but the trauma of travel has changed her cycle; maybe she appeared in season at home, but in fact she needed two more days before covering could take place. There are so many question marks that, unless you have a vet to scan the mare beforehand, you can never guarantee that she will stand. However, an older and experienced mare that you have known well for some years should not cause you problems.

In parts of Europe the owner may travel his in-season mare to the stud every two days during oestrus, or he may leave the mare at stud just for the period of her season. Certainly vet's fees at studs are often less than at home, due to the fact that the larger stud will have negotiated 'shared visit fees' for their customers. Therefore the decision whether to leave your mare at stud or to take her home after covering could be a financial one.

If the mare does not 'take' (meaning does not conceive) but you have left her at stud, there is at least in that case no travel involved and she is there, ready for covering again if necessary. If the mare does not return into season after 18–21 days, then scanning for pregnancy diagnosis can be carried out using the on-site stocks; and hopefully for you, the result will be positive.

If your mare does return in season, and she is not known to be particularly difficult to get in foal, then most studs will cover again, repeating the same every-other-day method as the first time. If she is unfortunate enough to return in season for a third time then the vet should almost certainly be called in to examine her. If your mare is younger or older than the recommended breeding age, then veterinary intervention is advisable, as it is for mares who are known to have been difficult to get in foal on a previous occasion.

Mares that are difficult to get in foal

Only the stud vet, with the aid of the scanner, can give reasons as to why a mare has not taken, and suggest various ways that may assist. It is not in the scope of this book to discuss the enormous number of conditions that may arise.

If a mare remains at stud (as opposed to being taken home after covering) the observation process will inevitably be more accurate and this will help the stud vet.

Mares that consistently succumb to uterine infection after a natural covering may benefit from the artificial insemination of fresh semen (AI), likewise an older mare whose cervix remains tight even when fully in season may benefit from artificial insemination (see p38). It is quite normal today for you to make a request for the stud vet to discuss your particular mare's problems. Remember that at most studs, every effort will be made to get your mare into foal, although there will always be some, for whatever reason, which do not conceive.

The shape of the mare's vulva is of vital importance to the success of fertility. For example if the vulva is long or sloping, there is a chance that infection may be set up because the vulva opening is particularly vulnerable to the passing of manure, and the sucking in of air during covering.

In older mares, especially those that have previously had foals, the shape of the vulva deteriorates, and it is often this poor shape that causes some mares difficulty in conception. The simple **Caslick operation** (named after an American vet) assists in eliminating the risk of infection. It involves stitching the upper parts of the vulva lips together, a process which is carried out by the vet under local anaesthetic, and in which a thin strip of skin is removed from each side of the vulva, the two sides then being stitched together. The vulva opening is therefore artificially reduced and this should eliminate unnecessary infection, as manure and air are now prevented from entering the vaginal area.

The Caslick operation: a local anaesthetic is given by the vet

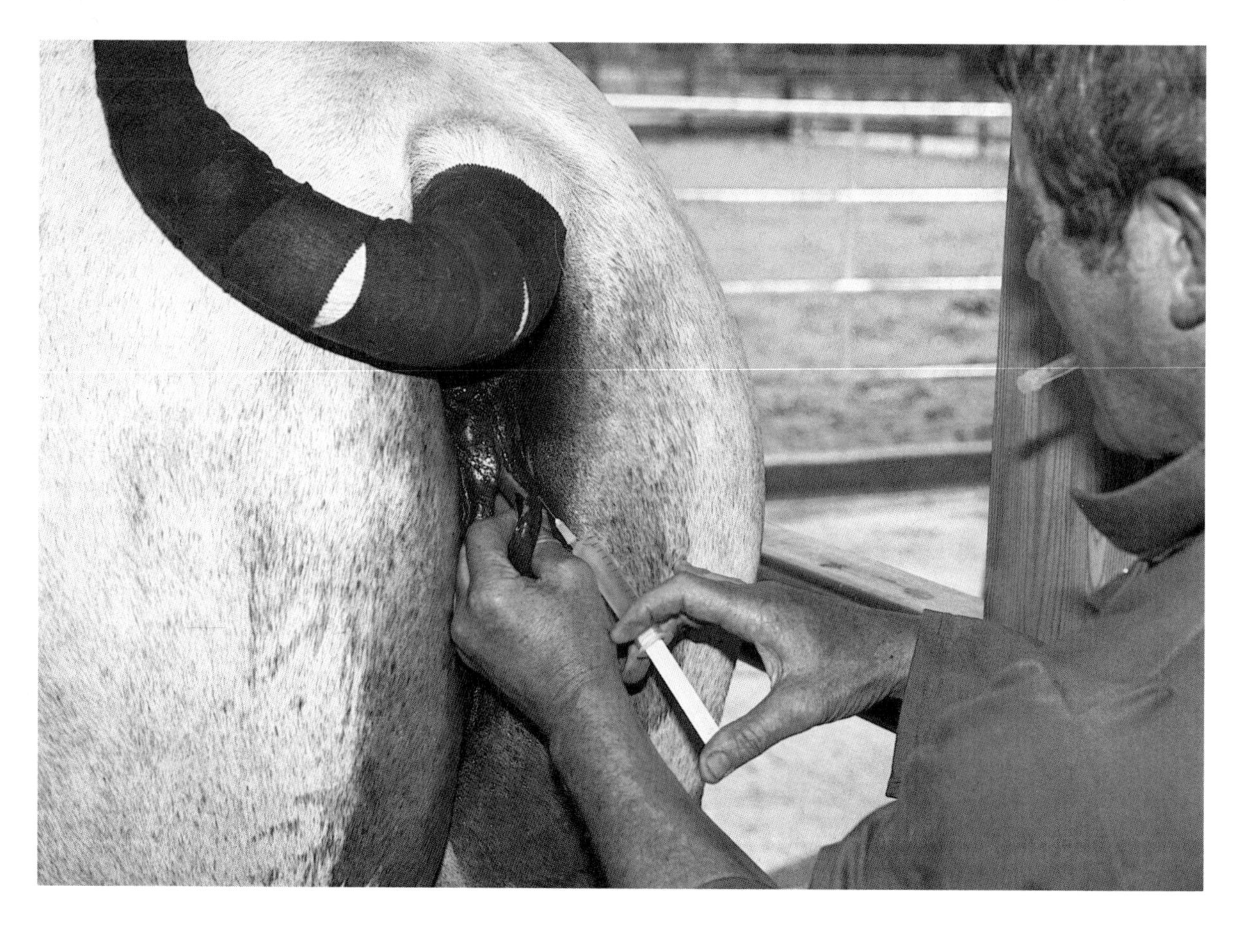

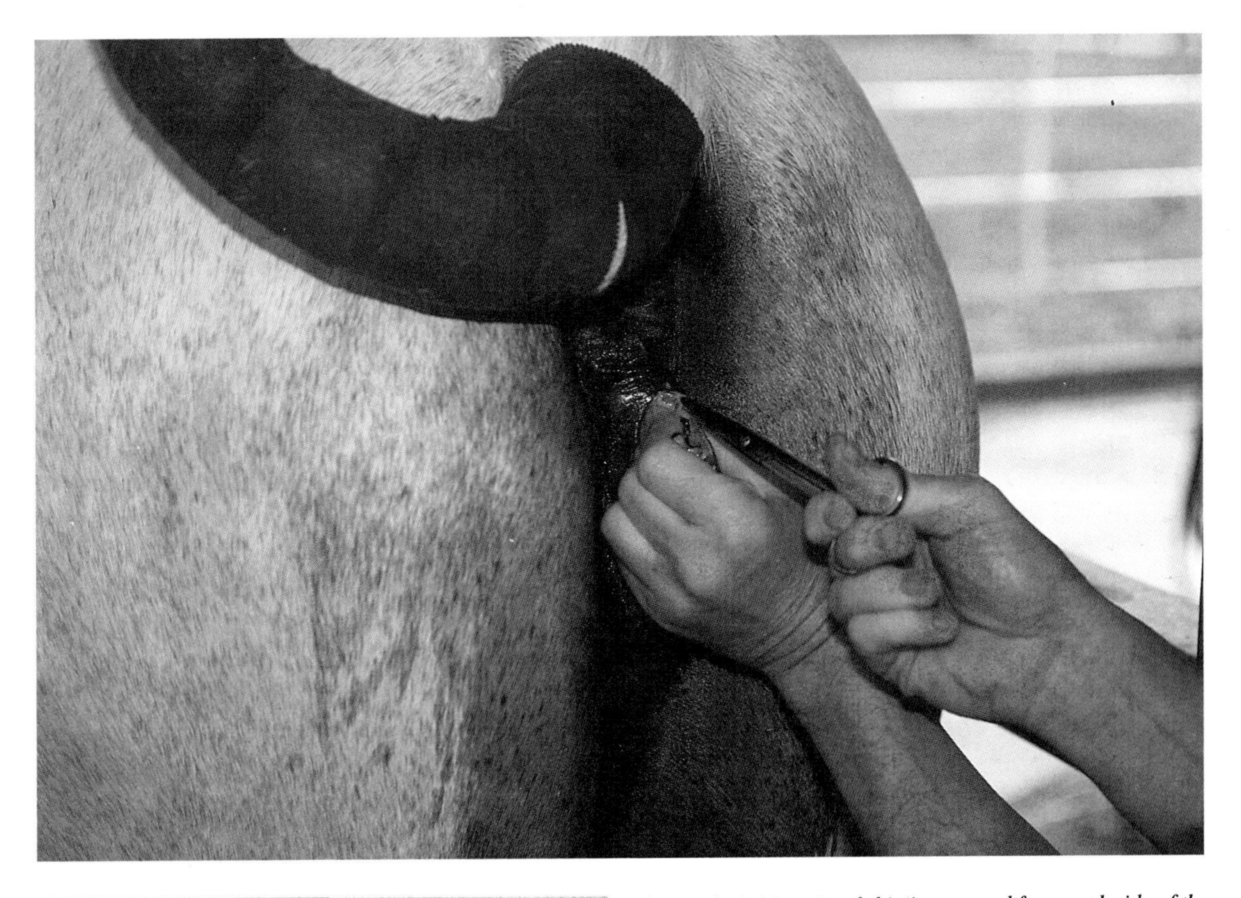

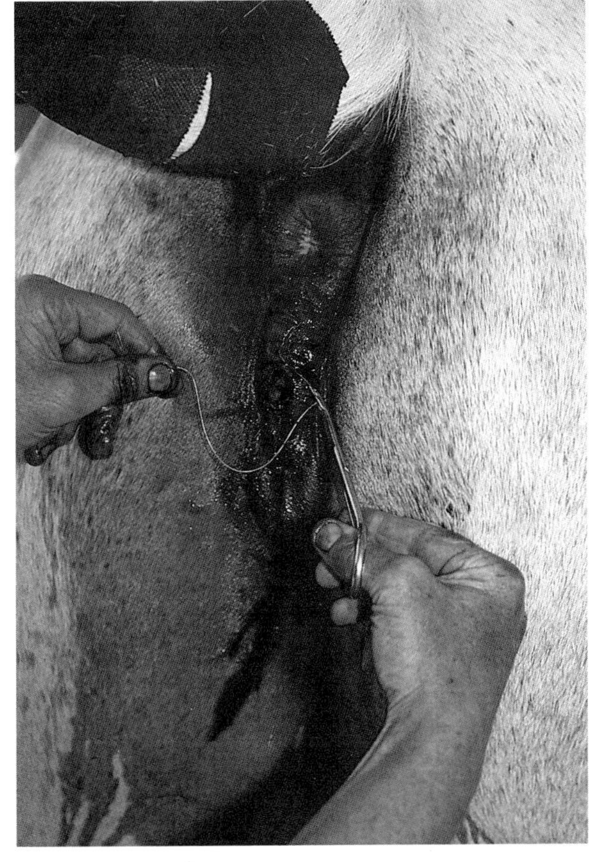

(Above) *A thin strip of skin is removed from each side of the vulva*

(Left) *The two sides of the vulva are then stitched up*

This simple operation is normally carried our immediately after covering (or in some cases three weeks later, when the mare is known to have conceived). It takes only a short time to perform and rarely needs subsequent attention other than to put some wound powder on the affected area for a couple of days. After healing, the actual stitches may be removed and a natural seal will form.

Most important: Remember that the area of the vulva which has been stitched must be cut open **prior** to foaling. As a result of having stitched the vulva there is only a small opening left – this allows urine to pass out, but there is **not enough room for a foal to be expelled**.

The opening can be cut approximately two to three weeks before foaling down – though at large studs when a vet is permanently on site and staff are **guaranteed** to be observing the mare at all times, the actual opening up of the vulva is done as the mare goes into the second part of labour. However, I recommend strongly that you call in your vet at least ten days before foaling, so that your mare doesn't go into labour without having had the vulva area re-opened. If she does, the lips of the vulva will be badly torn as the foal is expelled; such a jagged wound is slow to heal, and can deform the perineum. Alternatively, cut open the area yourself, using a new or sterilised razor blade. By the second stage of labour the vulva will be quite numb, so you will not be causing the mare undue pain.

ARTIFICIAL INSEMINATION

The use of AI as it is commonly known, was first introduced in Russian stud farms and later in Hannover, Germany, in the late 1940s; it is now an accepted breeding practice in non-Thoroughbred studs world-wide. Most countries have strict health requirements, regulations and methods of registration within the relevant breed society. The Thoroughbred breeding bodies do not recognise AI for reasons of policing and finance.

Artificial insemination is the collection from a male donor of semen which is then deposited into a receptive female to produce pregnancy. The technique of collecting, chilling and transporting chilled semen to the mare has been developed to such an extent in the last five years that it is now a viable option for breeding the competition horse. The use of frozen semen is also practised in the USA and Europe, but the fact that the semen of a stallion is very fragile makes it more difficult to deal with; currently in the UK mainly chilled semen is handled. If you are considering using semen from overseas, note that there are strict import regulations: check carefully beforehand that documentation is available and that the foal will be eligible for registration papers.

Fresh semen is collected from the stallion by using an artificial vagina (AV) and put into the

Equipment required for artificial insemination showing (from left to right) the artificial vagina used to collect the semen from the stallion, KY gel, semen extender, the Equitainer which is the cool box used to transport the semen, and a microscope

mare by the vet (inseminated). When both stallion and mare are at the same location the mare can be inseminated *immediately* after collection which is useful in the following cases:

a) when a mare will not accept a natural covering; or
b) when the stud vet advises that prepared semen may assist in conception.

Further advantages of using AI are that mare owners have a much greater choice of stallion, and distance is of no consequence. Thus a mare that does not travel well, has a sick or injured foal at foot, or will not or cannot accept a natural service, can now be accommodated; besides which the spread of any infection can be controlled.

One of the disadvantages of AI may be cost: although livery fees will not be applicable, there are veterinary fees (which can be high) and semen transport costs to be considered.

(Below) *A laboratory is essential on a stud where there are several stallions standing and artificial insemination is practised*

(Right) *The examination of semen under a microscope*

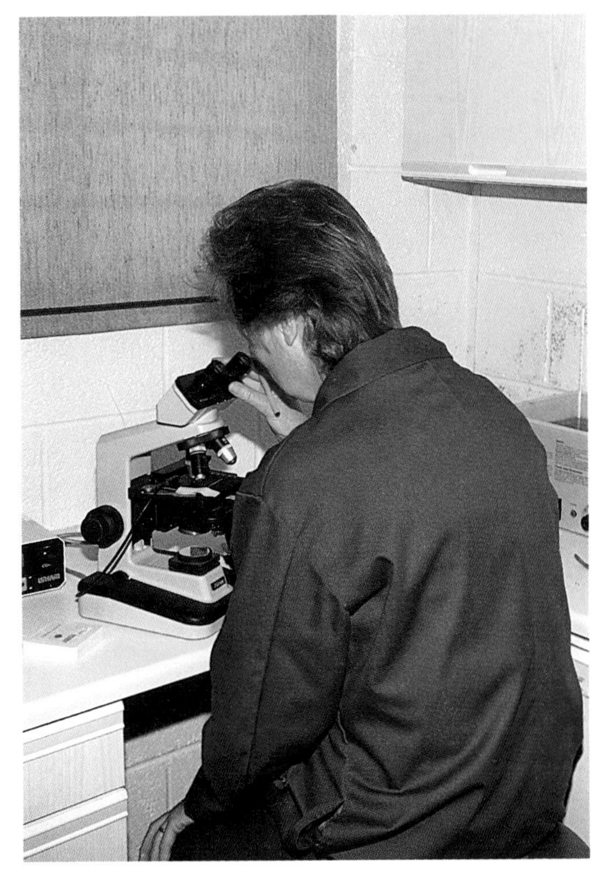

How AI works in practical terms, using chilled semen

If you have decided to use AI you need to act as follows:

1 Check that the stallion you have chosen is used for AI, and that the stud is familiar with the practice.

2 Contact your veterinary surgeon and check if he will carry out the insemination, or if he prefers to refer you to a colleague who is experienced in stud work.

3 Request that he come and examine your mare for breeding purposes, and make sure he takes the appropriate swab (see p22).

4 Complete the stud's nomination form (see p18-19) and send it back to the stud, together with a full description of the mare, the vet report on her fitness to breed, and the swab certificates.

5 Tease your mare with a stallion, rig or gelding so that you know when she is due in season.

6 Liaise with the stud and advise them of the appropriate date that you will need the semen. Also discuss how the container (that is, the receptacle in which the semen is dispatched) will be sent to you. In the UK the most commonly used container is known as an **Equitainer**, which maintains the semen at a constant cool temperature in an air-tight bottle. It will be sent by Data post or Red Star (UK) or a similar rapid transport method. Alternatively if you live close by, you could collect it, reducing costs further.

7 Once the mare is in season, request your vet for examination so that he can advise the optimum time for insemination. He will need to use a scanner for this, and may have to come out on more than one occasion. Not only does the semen have to be of good mobility, but the mare

Collecting semen from a stallion using a 'jump mare'; note the use of kicking boots on the mare's hind feet to protect the stallion if she lashes out at him

has to be at the optimum time during oestrus in order for fertilisation to take place.

8 Liaise well with the stud: this is important so that the stud knows when collection of semen is likely to be required. Most studs need a 'jump mare'. Some stallions can be trained to ejaculate into an artificial vagina (AV) while mounting a 'phantom', but it is more usual to use a 'jump mare' – that is, a placid mare in season, who will stand for the stallion to mount, at which time the semen will be collected. After collection the semen will be inspected under a microscope and subsequently prepared with semen extender for despatch.

9 The vet will advise you when to request the stud to despatch the semen (approximately 48 hours before ovulation). Note that it may be more difficult to arrange on Sundays, Mondays and Bank Holidays.

10 Upon receipt of the container (some studs prefer to despatch directly to the vet) insemination can take place. There should be enough semen for two inseminations (the second to take place the following day). In order to hasten ovulation the vet is likely to administer a luteinising hormone (LH) at the time of the first insemination.

11 A photocopy of the original nomination form will have been enclosed in the container alongside the verification of the semen: this must be completed by the vet and returned to the stud after insemination, to prevent any misconduct. The container will be sealed so the semen cannot be tampered with, and it is important to note that the actual insemination of mares must *only* be carried out by a veterinary surgeon, for reasons of safety, hygiene and professional knowledge.

12 Ultimately, the vet should inform the stud if pregnancy has been achieved so that a covering

SUMMARY: ARTIFICIAL INSEMINATION (AI)

1 **Fresh semen:** semen is introduced into the mare by the vet immediately after collection.
2 **Chilled semen:** the sperm is examined, evaluated and prepared for placement in a container. It is then rapidly despatched to the customer, and for optimum success should be used within 48 hours, although up to 72 hours is considered acceptable.
3 **Frozen semen:** is currently packaged in sealed straws, stored in liquid nitrogen. It has to remain frozen in clean, dry, dark surroundings that are not prone to temperature fluctuations. Due to the fact that a stallion's sperm is so very fragile and susceptible to damage, great care has to be taken in handling and preserving the semen. It may only be prepared at approved insemination centres. There are stringent rules and regulations appertaining to the preparation and storage of frozen semen.

The advantages of using fresh or chilled semen

1 Greater choice of stallions
2 Distance is of no consequence
3 Injured mares do not have to travel
4 Mares with injured or sick foals do not have to travel
5 Mares who will not accept a natural covering can be accommodated
6 Control of spread of infection
7 The possibility of establishing pregnancy in mares that often succumb to uterine infection following a natural mating
8 Continuous check on the stallion's fertility
9 No livery charges
10 Risk of mating accidents reduced.

The disadvantages of using chilled semen

1 The cost of the veterinary surgeon, due to the large degree of veterinary input.
2 The diligence required on the part of the mare owner, in liaising with the stud to ensure that the semen arrives when required. In short, the semen needs to be in the right place at the right time.

Conclusion

There is no doubt that AI using fresh and chilled semen is proving to be successful, and will gain more popularity with the increase of knowledge among veterinary surgeons and owners of both mares and stallions.

It is hoped that the present prohibitive costs of importing **frozen semen** together with the restricted import regulations and difficulty in maintaining the correct environmental conditions during import, may be overcome so that other countries can have easier access to some of the fine performance stallions from overseas.

certificate can be issued. If the mare did not conceive, the procedure has to be repeated.

Read carefully the **stud's terms appertaining to AI** – they may well be different to those relating to a natural covering, when the mare is 'on site'.

When your foal that was conceived by AI has been born, you will need to obtain the relevant registration documents. Blood samples will be required from both mare *and* foal in order for the breed society to have proof that the particular foal has the declared parentage. Breed societies already hold blood-typing of their listed stallions, but blood-typing is expensive – consult your vet.

EMBRYO TRANSFER

Embryo transfer in horses does work, and in recent years has met with considerable success. It is, however, an expensive exercise involving good equine management combined with the expertise of a top stud vet. **The process known as 'embryo transfer' works as follows**:

The fertilised egg of the donor mare is removed one week after ovulation by flushing out the embryo with a saline solution into a catheter. After examination it is transferred surgically (or non-surgically) into a surrogate mare whose oestrus pattern has to be synchronised with that of the donor mare. Ensuring that the donor mare and another mare cycle at the same time is extremely difficult even with the use of modern drugs, so it is advisable to use two or three potential surrogate mares in the hope that one will be in the required stage of oestrus to accept and continue the life of the embryo when it is transferred. Subsequent monitoring of the mare is required for some time.

Embryo transfer is advantageous where:

1 You have a top competition mare and would like to breed from her, but do not want to interfere with her competition schedule.

2 You have a mare who readily conceives, but consistently loses her foal prematurely.

3 You wish to increase the number of offspring of a rare bloodline or breed. In theory, six to eight embryos could be obtained from one mare during one breeding season, but in practice it is not so easy.

(Further reading on this subject can be found in specialist breeding books.)

THE DIAGNOSIS OF PREGNANCY

Whether your mare remains at stud or you take her home, a positive or negative pregnancy diagnosis (PD) needs to be established. There are a number of reasons why it makes sense to know the results sooner rather than later:

1 If you have taken your mare home, you will need to return her to stud if she is not in foal.

2 If she is still at stud and the result is positive, you can take her home and avoid further livery charges.

3 If the result is positive it is advisable to keep your mare separate from rigs and flirtatious geldings.

4 If twins are detected, a decision will have to be made between you and the stud vet as to what action is taken – see Twinning p44.

5 Depending on the mare's condition and the time of year, the feed may have to be altered.

6 It will help you plan ahead.

Methods of pregnancy diagnosis

Today the most usual form of PD is to use **an ultrasound scanner.** This scanning machine uses a pulse of high-frequency sound to contrast the organs of the abdomen and depict them on to the screen. The probe, which generates the sound pulses, is placed into the mare's rectum, and the uterus, filled with fluid, shows up as a dark mass with the embryo highlighted from within. The test can be carried out as early as 14 days or as late as 45 days (after this the form on the screen becomes too large to identify clearly). The recommended times are first at 18–21 days, and then at 25–35 days. This is the same method as is used for human beings.

Although pregnancy can be established using a scanner at 14 days, the foetus will only just show on the screen due to its small size, and a mare can easily reabsorb or lose the conceptus at this early

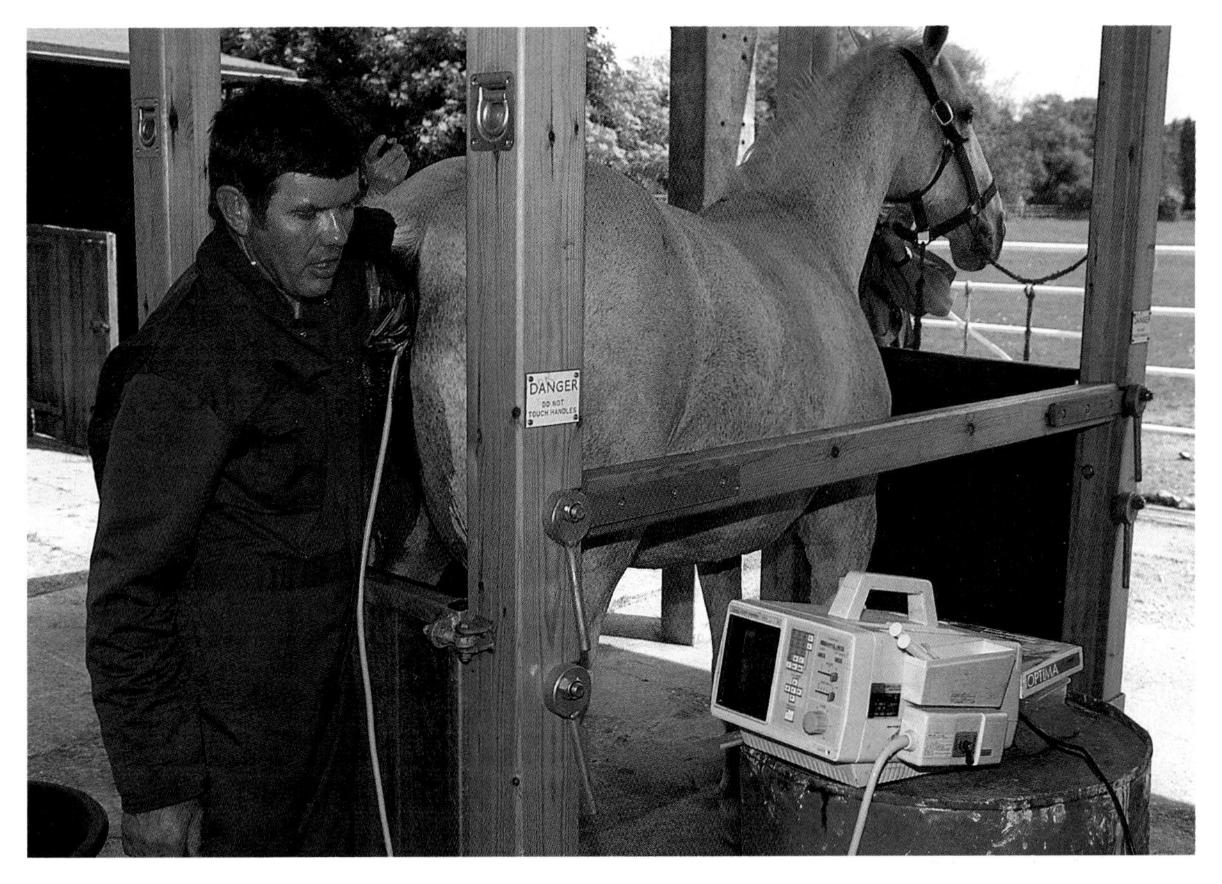

The stud vet scanning a mare for pregnancy diagnosis

stage. Twins, too, can be diagnosed but research has shown that one is very likely to be either reabsorbed or aborted during the early weeks of pregnancy (see 44).

If the mare remains at stud she will be presented to the stallion at about 16 days after the last covering of her previous season, and every second day thereafter to establish if she is going to return in season. If she does not, then there is every likelihood that she is in foal. Therefore scanning at 21 days is recommended in order to save unnecessary expense. Subsequent scanning at 25–35 days is also recommended to establish that the mare is still in foal, and by this time a heartbeat can usually be detected on the screen.

The scanner can also detect abnormalities (for example a very small foetus), and the picture of the uterus can portray other abnormalities to a good stud vet.

Polaroid photographs can be taken of the image of the foetus on the screen and given to the mare owner, if suitable equipment is available.

Manual pregnancy diagnosis

Until about three years ago manual pregnancy examination was the norm. The diagnosis was carried out at around 42–45 days in order to obtain the most accurate results. The veterinary surgeon would insert his gloved hand into the rectum and from that position could feel, through the rectum wall, if there was a swelling in one of the uterine horns. Today the manual method is still practised for a pregnancy diagnosis after 45 days, as the scanner is no longer so accurate at this stage in pregnancy.

Note: If your mare's nomination terms were **NFFR 1 October** and you are suspicious as to whether your mare is still in foal at the end of September, then you will need to call your vet to carry out a manual pregnancy examination to ascertain whether she has remained in foal. If she is in foal, all is well; if she is not, then it is your duty to send a vet's certificate to that effect to the stud in order to secure your free return the following season.

Blood-testing for pregnancy

This is not the most accurate test, as it relies on a specific hormone that is produced by groups of cells found in the placenta **between 40 to 60 days after the last service**. The hormone is known by the initials PMSG, yet mares who are not in foal could also have traces of PMSG. Kits to test for this are currently available on the market. Likewise another hormone, eCG is present in the bloodstream from between 45 and 100 days; a blood sample taken during this period and tested for the presence of eCG is used as a pregnancy diagnosis.

SUMMARY: ALTERNATIVE METHODS OF PREGNANCY DIAGNOSIS (PD)

1 The ultrasound scanner
First PD at 18–21 days
Second PD at 25–35 days (heartbeat can be seen).

2 The manual pregnancy test
First at 42 days
Second at end of September; reference 1 October terms only.

3 Blood-testing for pregnancy
First between 40 and 60 days – presence of PMSG
Second between 45 and 100 days – presence of eCG.
Both give *reasonably* accurate results.

TWINNING

Twinning occurs either because one follicle is fertilised by two sperms (causing identical twins), or because two follicles are fertilised resulting in two placentas (and unidentical twins). In all species other than horses, twins are carried successfully because there is space on the uterine wall for the attachment of more than one placenta. In the mare, however, **the placenta is designed for one foetus only**, and if two foetuses are present the space available is reduced by half, usually causing one or both to be reabsorbed or aborted. On many occasions the smaller twin is reabsorbed naturally, and the remaining embryo grows into a normal size foal. However, more than 50 per cent of twins are not aborted until 8–9 months into their term, and if one dies the other one is normally aborted; therefore **it is important to detect twins early on in pregnancy** so as to avoid this occurrence. If twins are detected when the mare is scanned at 18–21 days, the stud vet may leave nature to take its course and re-scan ten days later. At this stage (that is, at approximately 30 days), if twins are still apparent, he will make the decision either to abort both or, depending on the position of the two foetuses, to 'pinch' one out. There is a risk in leaving twins longer than 40 days, because if they are aborted after this time the mare may not conceive again for some months (due to the fact that eCG is produced at about this time).

If the two foetuses are very close, it is likely that the vet will abort them by simply administering a prostaglandin injection (PG) which will result in the mare coming back into season. If one foetus is 'pinched out', repeated examination by scanning will have to be carried out afterwards in order to check if the remaining foetus stays alive. If it does not, then the mare may need to be brought back into season with a PG. Twinning is known to be hereditary in some mares.

SUMMARY: TWINNING

1. If twins are sighted on the scanner at 18–21 days, the vet should decide whether to leave them and wait to see if nature takes its course and one foetus is naturally reabsorbed. If this is his decision, re-scan at about 30 days to check. If he decides to abort them, he will administer a PG and the mare will return in season within a few days.
2. If early scanning is not required, remember that if twins are left after 40 days and then aborted, your mare will be difficult to breed from for the next few months.
3. If one foetus is 'pinched out', re-scan to make sure that the other foetus lives. It may live for a week or two and then be reabsorbed.

PREGNANCY FAILURE

When abortion occurs before the end of the three-month term it is described as *foetal loss, embryonic death* or *resorption*. If a mare aborts before 150 days it is unlikely that you will find the foetus. It is either too small, eaten by a wild animal (if the mare is out at grass), or it was re-absorbed. You will only know that the mare has aborted if, after a while, her shape changes back to normal, or she suffers a vaginal discharge, or she comes into season. In all cases **you need to call your vet** to check the situation.

After 300 days, the loss of a foetus is described as *prematurity* or a *still-birth*.

At any time before 300 days the loss of a foal means that it has been aborted. In some cases, when the foetus and placenta are failing, the mare's udder will fill and she may run milk. However, if the foetus dies suddenly or the mare herself initiates the expulsion, then no milk will be produced. If an expelled foetus is discovered, then the vet should be called to check the mare and he is likely to recommend a post-mortem on the foetus. The reasons for abortion or still-born foals can be **infectious**, or **non-infectious**. Twinning (non-infectious) has already been described; other reasons can be found in more technical books on stud work, or in specialist veterinary publications.

3

THE GESTATION PERIOD

Your mare is in foal: those worries about sending her to stud, the technical terms that had to be added to your vocabulary, the concern that she may not conceive – what a relief, they are all over! But you are now entering the most important waiting game, the **gestation period** (the period of carrying the young in the womb between conception and birth). For a mare this is eleven months which is a long time, and keen observation, combined with common sense and efficient equine management are essential in order to maintain your mare in a happy, healthy condition. How should you manage your pregnant mare when she returns from stud? Should you keep her separate from geldings? Is it a good idea to continue to ride her, and if so, for how long a period? When should you have a further pregnancy diagnosis, or is that unnecessary? How do you recognise if there is anything wrong?

This chapter provides a simple step-by-step guide with illustrations of the mare's body changes and the growth of the foetus in the womb; how the embryo gets food and oxygen, and the major changes that take place in the body of the pregnant mare. Different methods of feeding will be examined, and it will be emphasised how important it is for the mare to take in sufficient vitamins and protein. To the human eye the gestation period is relatively dormant, but in real terms it is not, and if a mare is not in good condition, major problems could result as regards the future well-being of her foal and, in some cases, of the mare as well (see Case Studies p84 and p86). Therefore this chapter should be read carefully, and emphasis placed on the good management of your in-foal mare.

THE EARLY STAGES OF PREGNANCY

The foaling date

Once a mare is deemed in foal, the actual projected foaling date is always uppermost in a mare owner's mind. The full term of pregnancy is officially 340 days, although in practice it varies considerably and it is thought that the 'variable' is dictated to a certain extent by the mare's condition during her gestation period (see p61-2).

In the wild the mare is often observed to go a complete year before giving birth, allowing the foal more time to grow to its required weight before giving birth. However, this may be because she has only poor quality pasture – a mare that has been 'well done' in terms of protein and the relevant vitamin supplements is more likely to foal on or around the 340th day.

On return from stud

During the period after your mare returns from stud she should be no trouble. She will be able to remain out at grass until the autumn and, provided that a good supply of clean hay is fed as the grass begins to wane, and that the early autumn is not too wet, she should be happy to graze in her new-found capacity as a brood mare. Good grass is the most natural and complete food.

It is important that you consider carefully what company your mare goes out to graze with. The question is always raised, does it harm to put a pregnant mare with a gelding? Basically if a mare is confirmed in foal, in a similar way to when a woman is pregnant, she is not likely to lose her foal. However, it does happen and it is fact that **during the early stages of pregnancy the mare is more vulnerable**. Certainly a mare can lose her foal as a result of continuous aggravation by a flirtatious rig or gelding. A 'rig' is the name given to a male horse who has one or both testes still in his abdomen (undescended); a gelding is said to be 'riggy' when he behaves in a stallion-like manner. They may not be in the same field, but even geldings in adjacent fields can cause problems. In short, **do not subject your in-foal mare to stallions, rigs or flirtatious geldings,** especially in the early stages of pregnancy.

The critical period is between 5 and 8 weeks into pregnancy, due to the fact that the anchorage of the conceptus to the lining of the uterus could easily be weakened – because of this the mare's system is especially susceptible to abortion. Sudden changes in food, environment and companions should be avoided, as should upsetting situations which result in excitement and sweating.

A brood mare does best in tranquil surroundings – a busy livery yard where stabled horses are put out into the same field as your in-foal mare and gallop around for five minutes each day is not to be recommended. If you do not have ideal facilities at this stage, it may be preferable to leave the mare at stud for a while longer.

Riding the in-foal mare

There are various schools of thought on this subject, but in the days when the horse was used as a work horse the mare continued to work while carrying her foal, often up to one month before foaling. There should be no harm caused by continued gentle riding of the mare, *but* to subject her to excitement, short bursts of galloping and jumping is *not recommended*. A mare with a calm temperament is the ideal candidate to continue light hacking, *especially* if she is prone to putting on weight, such as a pony mare. In this way you will keep her trim, as an in-foal mare should not carry excess fat – being over-weight can cause a mare to abort.

How long you continue to ride is up to you – use your common sense. However, after 5 months it must be uncomfortable for the mare to carry a rider's weight as well as the growing foal, and at this stage her muscles need to relax, allowing maximum space for the growth of the foal (see Case Study 1 p84).

Worming and blacksmith's visits

During this period the **normal worming programme** should continue (at least once every 8 weeks), **as well as regular attention to the mare's feet** (every 6–8 weeks) – if she has been put out to pasture as a brood mare you may well have had her shoes taken off, but the foot growth rate is faster during the spring and summer months. This, combined with extra weight being carried by the legs and feet, means regular paring and trimming should not be overlooked. Special attention should be paid to mares who have flat feet.

Teeth

The teeth are often forgotten: make sure they are not sharp and that the mare can masticate well – sharp points make chewing painful and so reduce digestive efficiency. I recommend a visit by a specialist horse dentist on an annual basis, and you may find that the teeth of an older mare need rasping every six months.

Vaccinations (USA)

In the United States a mare should receive, as a matter of course, three rhinopneumonitis vaccinations during her pregnancy, in the fifth, seventh, and ninth months. A mare left unvaccinated is susceptible to the equine herpes virus that can cause equine abortion. These vaccinations are available in the UK, and their use should be discussed with your vet.

Four to six weeks before the mare's due date she should receive booster vaccinations as prescribed by your veterinarian. In the United States, these vaccines usually include Eastern and Western encephalomyelitis, tetanus, and influenza. At this stage of the pregnancy the immunity will be passed to the newborn foal via the colostrum.

THE GROWTH PROCESS OF AN UNBORN FOAL

Unlike most domestic pets, the foal remains inside the mare until an extremely advanced state of development is reached. The term **embryo** is the name used to describe any period in the stages of development from a fertilised egg to an animal ready to be expelled from its mother's womb, *but* when an embryo has developed sufficiently to be recognised as a foal it is often known as a **foetus**.

Life in the womb can be studied in greater detail in veterinary manuals, but other points that are interesting to note (in readiness for comprehension when birth takes place) are as follows.

Very strong blood vessels run from the foal's heart to its lungs ready to take the enormous supply of blood which is needed by the lungs when it takes its first breath of air (see p73).

Oxygen and food are taken in by the mare and eventually passed to the foetus through the blood supply, hence the importance of good feeding (see p51). Although no solid food is available, it is thought that as the digestive system develops the foetal foal may swallow amniotic fluid (the fluid in the placenta in which the foetus is suspended) from time to time, which in effect exercises the canal from the mouth to the stomach as well as the stomach itself and intestines. **This prepares the foal for suckling and sets up the important 'sucking reflex'**.

Movement of the foetal body should begin a few weeks before foaling time, and violent kicking of the limbs may be noticed during periods of observation.

Physical change in the mare

During the early stages of pregnancy the change in the outside appearance of the mare is minimal, although as you can see from the step-by-step diagrams of the foetus's growth, a series of complicated changes take place in the mare's body. However, the actual size and weight of the foetus do not increase dramatically, so the abdomen shape and size does not visibly alter until about four months. If your mare is a maiden, the foal will be even slower to show, due to the fact that she has a 'tighter form', not having carried a foal previously. A regular breeding mare may show much sooner and will, over the years, retain a 'dropped' abdomen unless she is put back into work at some period between breeding.

1 October terms

It is important to remember that if you have any doubt that your mare is in foal, and the stud terms applicable were NFFR 1 October (No Foal Free Return 1 October), then you should arrange for a pregnancy diagnosis so that if she is not in foal

Step-by-step diagrams of the growth of the foetus

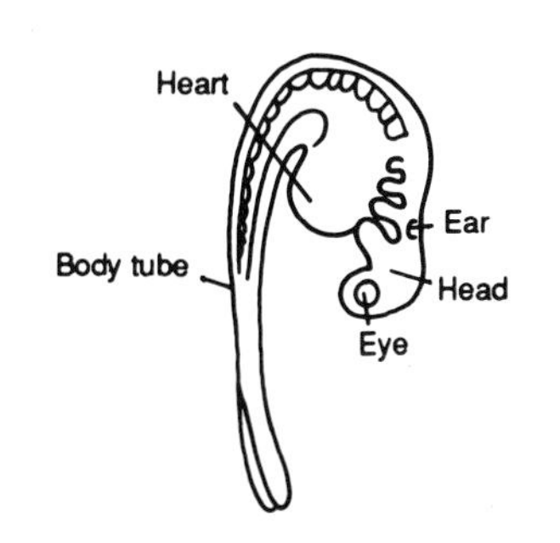

About 25 days (3 weeks plus). *Immediately after fertilisation has taken place the previously dormant egg becomes very active, and by the third week the beginnings of the future foal can be observed*

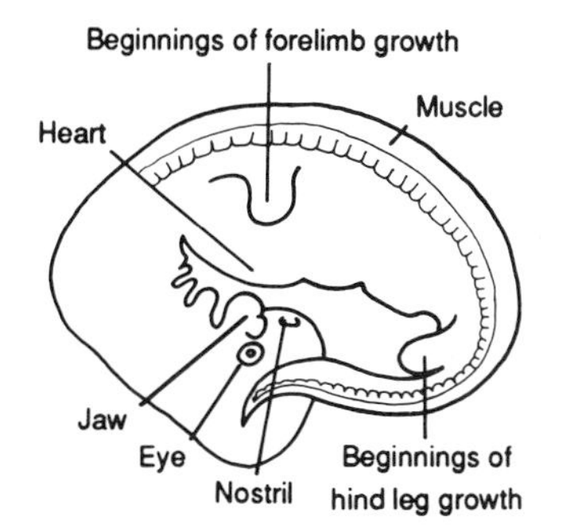

About 35 days (5 weeks). *The embryo now has the following: a large brain, a heartbeat, lungs, larynx, mammary glands and ear flaps. About 1cm long; weight 0.1g (horse foal)*

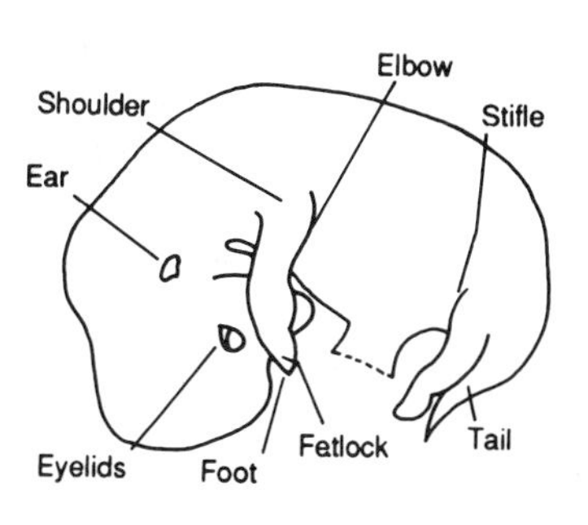

About 6 weeks. *The embryo develops traits that make it begin to be distinguishable as a horse*

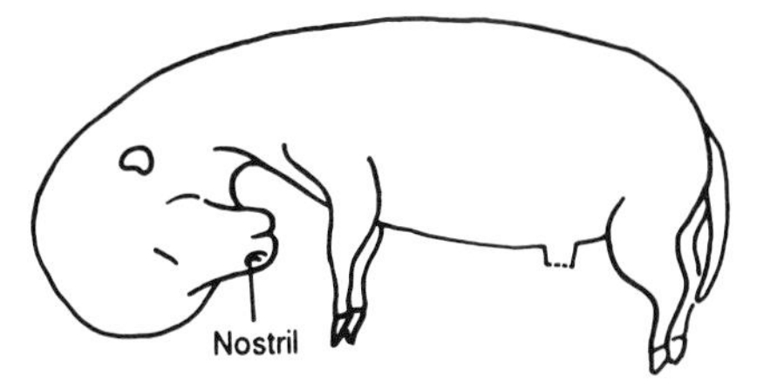

By 8 weeks *the foetus becomes horse-like in profile – now approx 5cm long. Nostrils are apparent, eyes and eyelids have formed, and the neck is longer. The stifle, hock and knee are recognisable. Ear flaps cover the ear canal opening. Fetlock and hoof are formed, and the sex of the foetus can be determined*

About 4 months. *By now all the internal organs are well established and from this time on the foetus grows in size and weight as well as specific detailed structural growth. About 25cm long, the weight is approximately 1kg (horse foal)*

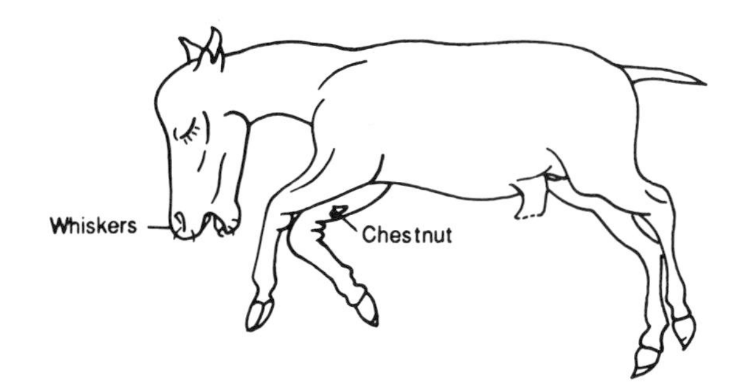

About 6 months. *By this time the foetus looks quite similar to an adult horse. The finer details are developing, including the hair for the mane and tail, ergots and chestnuts, plus the hair around the muzzle and chin. The limbs are now growing faster than the rest of the body*

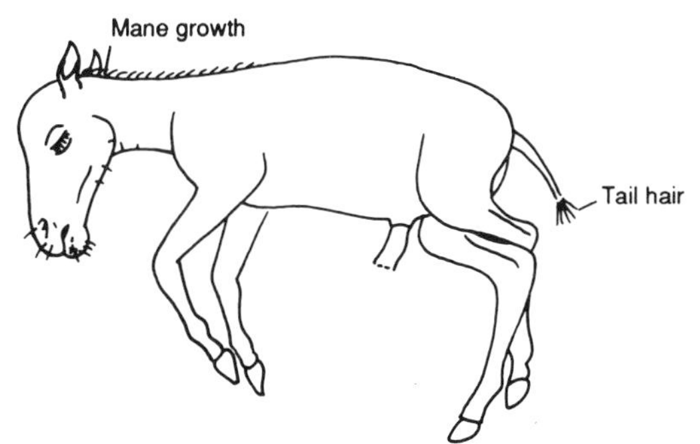

9 months. *In the final two months there is great increase in size and weight in preparation for expulsion; during the last month the foetus gains 3/4lb (1/3kg) per day. In addition, body hair and the hair of the tail develop, and a pad of soft horn is formed on the feet so that the womb is not damaged during birth*

(Above) *A ten months pregnant maiden mare; note that her stomach is not 'dropped' as in older brood mares*

(Right) *Ten months pregnant brood mare*

you can send the certificate to the stud in order to ensure a free return the next season.

THE LATER STAGES OF PREGNANCY

Assuming your mare was covered during the spring or early summer months, then a little hay should be fed in early autumn as a supplement to grass, and by October it is likely that you will need to feed a little hard feed as well. You may well have to bring your mare in to shelter at this time. Many studs today, especially in Europe, use a barn for their brood mares in the worst of the winter months; this cuts down the labour costs as they can be kept on a deep-litter system. Of course, for one brood mare alone this is not practical.

Suggested feed charts can be found on p53.

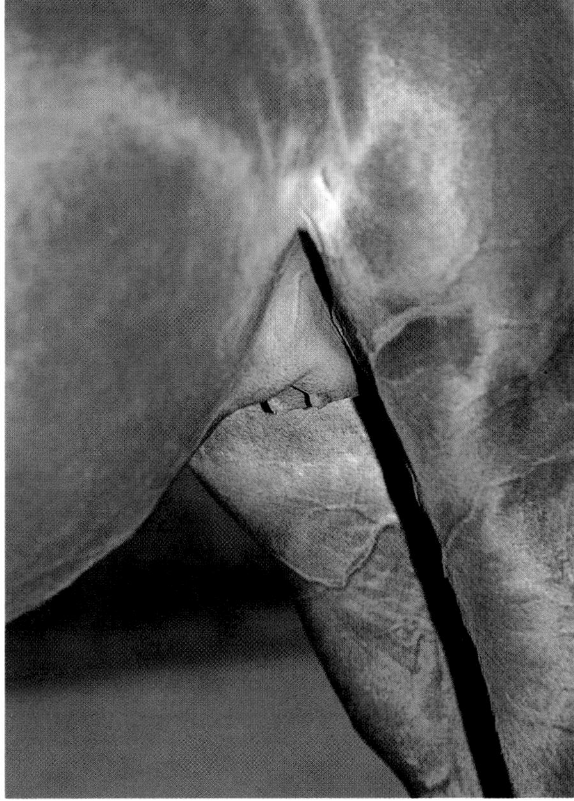

(Top) *Eleven months pregnant* overweight *pony brood mare; it is important to monitor a brood mare's weight in the early stages of pregnancy*

(Above) *The first signs of bagging up. A maiden mare's udder five weeks before her foaling date*

Remember that feeding 'little and often' applies even more to the brood mare due to the fact that her abdomen is becoming larger all the time. **One large feed per day is not advisable**; however she should do well on two feeds, providing she is out at grass on a daily basis. The brood mare needs her exercise. During **the last three to four months** it is important to feed additional protein to ensure correct growth of the foal, supplemented by a recommended vitamin with the correct ratio of calcium to phosphorus.

The breed, type and size of your mare will dictate how much feed is required, but **good observation** on your part is vital. If your mare is well covered, has a shiny coat and is bright in herself, then all should be well. If she looks fat, or is listless, or her ribs show, then take different action. Often during the latter stages of pregnancy the mare becomes a fussy feeder and may leave some of her food. Providing your observations are correct and she looks good, do not worry: it is Nature's way of saying that the intestines cannot be overloaded with bulk at this time. Succulents and weekly Epsom salts will help to prevent constipation.

During the last four to six weeks the 'bag' (udder) will begin to spring (develop), especially during the night when the mare is resting; it will reduce again in size during exercise, but towards the last week or ten days prior to foaling it will remain swollen during the day as well (see p62).

Feeding the pregnant mare

Most horse people are aware of the golden rules of feeding, and their application is certainly important when feeding the in-foal mare. Likewise common sense will always play a vital part.

For years, feed for horses simply consisted of oats, hay and grass, but today it is far more technical; scientifically researched compound feed is accepted, even by those whose knowledge and ability to feed by 'eye' has been handed down over generations. Much time, money and research has been put into the new approach to feeding, and 'complete feeds' are certainly one answer: the ingredients are guaranteed, as is the consistent quality. For the mare owner **stud cubes**

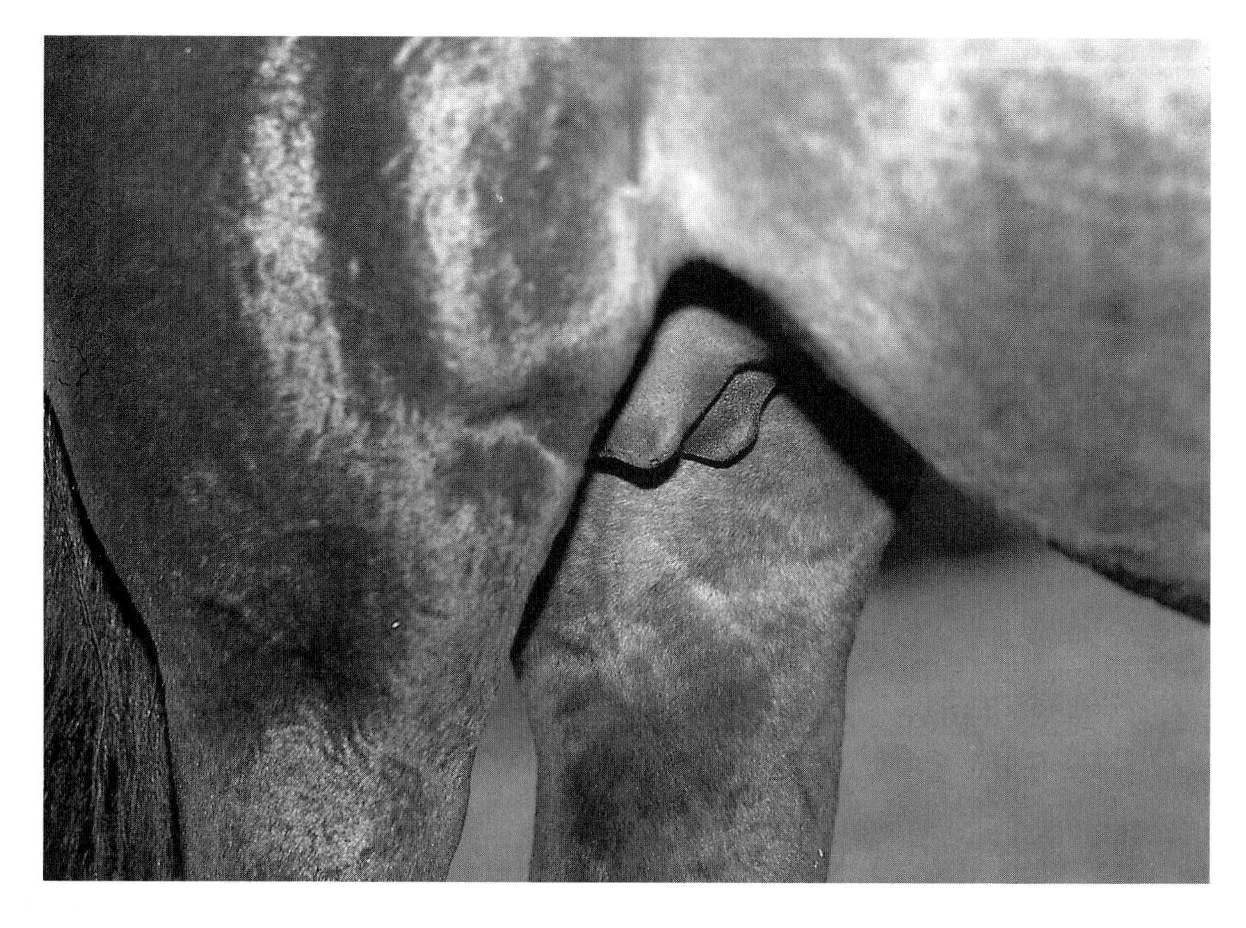

A brood mare's large teats which should not be mistaken for swollen udders

or **stud diet** have been designed for breeding mares, who need **quality proteins** to provide **adequate energy and micronutrient levels** for foetal growth and subsequent milk production. The balance of nutrients typical in a bag of feed might be:
Protein 15% (150g/kg); Oil 3%; Fibre 9.3%; Calcium 1.2%; Phosphorus 0.7%. A detailed breakdown of feed content can be found on every bag of feed.

It is important to ensure an adequate supply of minerals throughout pregnancy; in the pregnant mare (and all youngstock), the ratio of calcium to phosphorus is vital to ensure correct bone growth and development. Whatever dietary brand you choose, the company should offer a vitamin supplement that complements their particular feed. Some years ago calcium was supplemented by using bone flour, but since cattle have had problems with diseased bone, it is preferable to use a proprietary brand of vitamins with a greater amount of calcium and phosphorus, or to add **limestone flour**.

Even when the mare does not require hard feed during the gestation period, it is still advisa-

IMPORTANCE OF CORRECT FEEDING DURING PREGNANCY

1. Do not change any feed rations suddenly. Change over a period of 7–10 days.
2. Ensure that the mare is put out to graze on a daily basis – grass is her natural feed.
3. Several small feeds are nutritionally more beneficial than one large feed.
4. Well in-foal mares will have difficulty in digesting large feeds with a lot of bulk.
5. Endeavour to feed at regular times which will aid the digestion process.
6. Do not cut out all protein, even if your mare looks very overweight. Remember the mare is feeding the foetus as well as herself.
7. Give extra minerals by feeding a suitable vitamin supplement, especially in the late stages of pregnancy.

ble to feed a suitable vitamin supplement, as a grass/hay diet *does not* always contain sufficient calcium and phosphorus. **As pregnancy progresses the mare needs a gradual increase of protein,** and it should be noted that feeding a high protein feed to a well in-foal mare is unlikely to make her over-active and sensitive even if previously she reacted in a dizzy manner to such a feed. **Brood mares need protein to ensure the correct tissue growth of the foal**; for example, the resting adult horse requires 10 per cent crude protein in the diet, but during the last three months of pregnancy a mare needs about 16 per cent. **Reduce bulk in the last few weeks of pregnancy but maintain the protein level.**

In the last few days before foaling, your mare should be on a laxative diet: add some Epsom salts with bran (at the current time a controversial feed stuff, but in this instance should do no harm). If you can estimate the time correctly, simply feed hay and low energy feed hours before birth (see p61).

TYPICAL EXAMPLE OF A STUD FARM FEED CHART NOTED IN NOVEMBER

Assume mares were covered in May and are six months pregnant. Feeding of hard feed began at the beginning of October and has gradually increased.

Never feed without knowing how many pounds of feed each scoop is worth – for example stud cubes will weigh more than stud diet and oats. For the chart below, assume feeding stud diet when the scoop equals 2lb of feed. Hay and sugar beet should be fed as necessary and according to the condition and build of the horse or pony.

Note: These quantities may not be suitable for ponies and small horses prone to laminitis. Check with your vet.

Brood mares	**Pony**	**Lightweight cob**	**Horse**	**Large horse**	**Vitamins**
am	Hay	2lb stud diet Sugar beet	3lb stud diet Sugar beet	3lb stud diet Sugar beet	
pm	Hay	2lb stud diet Sugar beet	3lb stud diet Sugar beet	4½lb stud diet Sugar beet	Yes

TYPICAL EXAMPLE OF A STUD FARM FEED CHART NOTED IN JANUARY

Note: Hard feed and therefore protein levels have been increased.

Brood mares	**Pony**	**Lightweight cob**	**Horse**	**Large horse**	**Vitamins**
am	½lb stud diet	3lb stud diet Sugar beet	3lb stud diet Sugar beet	4lb stud diet Sugar beet	
pm	½lb stud diet	3lb stud diet Sugar beet	4lb stud diet Sugar beet	5lb stud diet Sugar beet	Yes

4

FOALING DOWN

The sight of a newborn foal suckling from your mare is beautiful: to go into the yard in the morning and find her standing over a live healthy foal, or to check the field on a midsummer day and discover that she has given birth, the foal is drinking and the afterbirth has been expelled, brings a great feeling of relief – what a lucky owner you are! However, many stories are told of lost foals and sometimes even lost mares, perhaps due to negligence, poor stable management, lack of preparation, or just sheer bad luck.

This chapter intends to try and help prepare you for the course of events immediately prior to foaling and during the actual birth – by knowing what to expect many problems can be eliminated. The first decision to make is, 'Where is your mare going to foal down?'. What facilities and equipment are essential? Will you be able to keep your mare at home, and when foaling is imminent, will you have the time and energy to sit up at night? Are you sure you know the external signs that point to imminent birth? Do you know when to call the vet? Is there going to be someone at hand during the day, in case the mare decides to foal down outside in the pouring rain? Would it be better to send her to a stud whose knowledgeable staff you feel you can trust? . . . but then you would miss out on the actual birth.

These are points that you, as the owner of a pregnant mare, need to consider very carefully. After all, you have waited nearly a year for this event to take place and have probably spent a lot of money in the process of getting your mare in foal. In the following pages you will find a guide to the basic equipment recommended for foaling down at home, a detailed account of the actual birth and how you can help, guidance on the care and observation of your mare and foal immediately after birth, and most important of all, when to call for outside assistance.

A purpose-built foaling box with boxed-in water bowl and feed manger

PREPARATIONS FOR BIRTH

Where is your mare going to foal down?

Initially most owners intend to foal down their mare at home, and it is only when they consider the responsibility that they perhaps change their minds. In order to assist you in making this important decision, I shall indicate here what would be the ideal facilities required at home:

1 A **good size foaling box** with a top door: for a 16hh mare 14ft x 14ft (4.25m x 4.25m), and for a 15hh mare 12ft x 12ft (3.70m x 3.70m). The box should have no protruding objects or rough walls, and should be near enough to running water so that it can be meticulously cleaned and disinfected before foaling. Remember – if you have to use your mare's normal stable for foaling down, then provision should be made to use an alternative stable for the few days it takes to clean up the proposed foaling box. (Do this approximately four weeks prior to foaling.)

2 Electricity should preferably be on site; if there are any problems at night the vet will then be able to see to work without hindrance. However, a good torch is quite adequate.

3 Access to hot water which could be needed by the vet.

4 A suitable paddock to put out your mare and newly born foal – this needs to be close to the foaling box. Ideally the paddock fence should be post and rail or hedging. Barbed wire or wire fencing are totally unsuitable for young foals; however, electric fencing, if properly maintained, can be effective and should cause no harm. The field should be relatively flat, that is, not on a steep slope or hilly ground, as bone-growth problems can occur from overstrain on the limbs. *Make sure there are no surrounding ditches.*

There is a possibility that your mare may foal outside, which is fine providing the field is safe and easily observed from your yard or house so that action can be taken quickly should there be a problem. Mares always seem to foal in an awkward spot and a ditch could mean real trouble if, for example, the mare gives birth at the edge and the foal rolls into it.

5 It is advisable to have somewhere you can easily observe from at night, eg an adjoining tack room or nearby caravan. Or even better, video or foaling alarm facilities.

6 Make sure that your local vet is aware that your mare is going to foal down, and check that he is willing to attend at any time of the night in case there are complications.

SUMMARY: FACILITIES REQUIRED FOR FOALING DOWN AT HOME

1. Foaling box
2. Suitable turn-out paddock
3. Observation point
4. Local vet willing to attend
5. The telephone number of a nearby stud in case colostrum is needed urgently
6. The telephone number of your local feed merchant who stocks mare's milk replacer
7. The telephone number of the National Foaling Bank
8. Foaling bucket

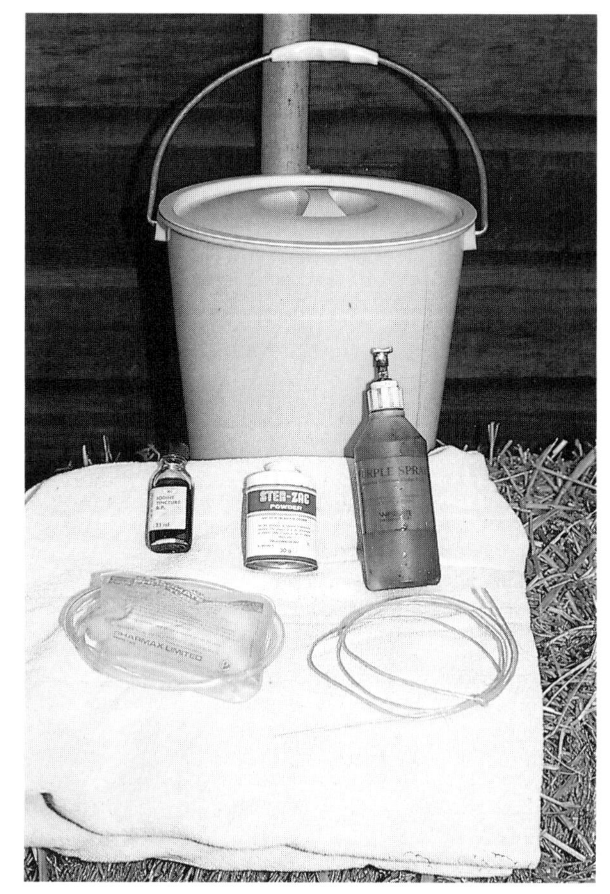

The foaling bucket and contents showing (from left to right) iodine, antibiotic powder (in this case Ster-zac), purple antiseptic spray, enema to use on the foal immediately after birth if necessary, and string to tie up the afterbirth

Once you have decided that the yard facilities are suitable, and have made arrangements with the owner if the premises belong to someone else, then you have to make the personal commitment to be on the spot and available to observe at the critical time. All this is, of course, made much easier if you have a friend or colleague who has witnessed foaling before – at least they know what to expect – and who will presumably share the sitting-up at night when birth is imminent.

Taking your mare to stud for foaling

Alternatively you could send your mare to stud for this important event. **The advantages are as follows:**

1 You know the facilities are good – you have been to inspect.

2 There are knowledgeable staff on site, and all the necessary foal equipment is in situ.

3 You will not have to sit up and observe at night time prior to foaling.

4 Your mare is probably in a more suitable environment should there be complications.

5 If you intend putting your mare back in foal there will be no worries about transporting her and the newborn to stud (unless you decide to use AI – see p38).

The disadvantages are:

1 It is undoubtedly an expensive exercise.

2 Even though you would, I am sure, always be welcome to visit, you are unlikely to witness the moment of birth (although often that is also the case at home).

Many people feel that it is infinitely more preferable to allow your mare to foal at home: she will feel more settled and relaxed with those she knows, and altogether happier in her own environment.

If you decide to take your mare to stud for foaling, you will need to discuss at what point you are going to send her. Most studs will suggest she

arrives **approximately six weeks prior to foaling**, and a minimum of four weeks.
Reasons:
1 She needs to build up her own antibodies in the new environment.
2 She needs time to settle down before foaling is imminent.
3 It is better for your mare not to travel in the last month of pregnancy.
4 The stud staff need to become familiar with her behavioural patterns.

Foaling down at home

The decision is made – your mare is going to foal down at home. At least six weeks before the date you should gather together the following equipment; the items listed below are used at every birth. These are the minimum requirement for your foaling bucket. Some studs prefer to use purple antibiotic spray instead of iodine or Ster-zac powder. Iodine tincture and Ster-zac powder must be used in conjunction.

Foaling bucket	Application	Available from
1 Plastic bucket with lid, to contain items listed below	Used to keep equipment sterile and free from dust	Household store, baby store
2 One 25ml bottle *iodine tincture BP*	½ to ¾ bottle applied to the foal's umbilical cord stump immediately after birth	Local drug store or chemist
3 *Ster-zac powder* If no **Ster-zac** at hand use any wound powder *but* it may not contain the same proven ingredients	Applied to the foal's umbilical cord stump (navel) straight after the iodine to dry up any wetness	Local drug store or chemist
4 Cotton wool	a) To protect the skin from burning when iodine is applied to the navel b) General veterinary purposes	Local drug store or chemist, or saddler
5 Surgical gloves	Used by the person who applies items 2 and 3 to the foal's navel and useful for other occasions	Local drug store or chemist, or vet

Further useful equipment it is recommended to have on hand at all times:

Equipment	Application	Available from
1 Enema	For use if foal retains meconium or straining occurs	Vet, who will advise how to use it
2 Feeding bottle and lamb's teat with holes opened up	This will be needed in the following cases: a) if the mare dies giving birth b) if the foal does not suck after birth – – see p78 c) if supplementary feeding is recommended by the vet	In UK only: *The National Foaling Bank; *Horse Requisites, Newmarket. Otherwise, in an emergency purchase a baby's bottle from the local drug store or chemist, or large supermarket – cut a larger hole in the teat
3 Plastic jug (1 or 2 litre size)	To collect colostrum or milk from the mare's udder in cases when the foal will not suckle (or when colostrum is required for freezing, see p79)	Large supermarket
4 Mare's milk replacer	Used in cases of emergency: a) if mare dies at birth b) milk dries up c) supplementary milk is required. If you do not wish to purchase this in advance, make sure you know where to obtain it (most studs have mare's milk replacer on site)	Local saddler, feed merchant, manufacturer or stud farm

Useful items to add to your normal veterinary stock:

Equipment	Application	Available from
1 Liquid paraffin	An oral dose of liquid paraffin may be needed if the foal is seen to strain after birth (and an enema has not released the meconium), or later in the foal's life, when an enema is not necessarily applicable	Vet, drug store or chemist
2 Kaolin & morphine	For oral use when foal scours	Drug store or chemist
3 20cc syringe	For oral doses of items 1 and 2 above	Vet

Approximately four weeks before the mare is expected to foal down, you should gather together the following equipment:

Item	Application
1 Headcollar and tail bandage	To put on mare as she begins to foal
2 Large clean bath towel	For rubbing down newly born foal
3 Soap and hand towel	For use by vet
4 Two overalls	To be worn by handlers at time of birth
5 Jute, string or baler twine (nylon string tends to slip)	To tie up the afterbirth
6 Dustbin liner	To put the afterbirth into for subsequent burning or burying
7 Torch	In case extra light is required
8 *Your vet's telephone number*	In case of need
9 The champagne bottle	For celebration purposes!

The Foaling Box

Approximately four weeks prior to the foaling date you should prepare the mare's box. Naturally, if the box is to be used solely for this purpose then preparations for cleaning and disinfecting can be made some time before; once the stable is dry, the top door should be closed. Remember to keep out dogs and cats.

Three to four weeks before the date, as your mare begins to bag up well, put down a **good deep bed of clean (dust-free) straw**. The walls should be banked up about 3ft (1m) high and thick so that when the mare is restless they retain their solidness. The corners should be well packed and if there is an automatic drinker or manger, the straw should be banked right up underneath, covering all naked pipes or taps. Plastic corner mangers in frames should be fixed on to the frame with string (if a temporary measure) to avoid the danger of the mare kicking the manger out and leaving a gap that she could get her legs under when rolling. Ideally the frame should be encased.

Never use a hay-net – they are dangerous for mare and foal – put the hay on the floor. *Do not* stand buckets of water on the floor – the foal could drown at birth. I recommend you purchase bucket clips from your local saddler, but make sure that they are high enough so that there is no danger of the mare knocking the bucket off the wall when in labour. If feeding from the floor, remove the handle from the bucket as it could be dangerous; preferably use a Kanguroo feed bucket made of thick, heavy rubber which has no metal handle, or feed from a manger over the stable door.

The foaling box plays a large role in the safety of foaling down. Follow these instructions carefully, and you will have eliminated some of the accidents that can occur.

SUMMARY:
FOALING BOX

1. Clean and **disinfect thoroughly** both walls and floor
2. Use good quality, clean, dust-free straw
3. Bank up bed well under manger and water drinker
4. Fix plastic feed mangers to their frames
5. Do not leave feed buckets with handles on the floor
6. If no automatic water drinkers, fix water buckets safely to the wall
7. Do not use a hay-net
8. For hygienic purposes, keep dogs and other animals out of the stable

The foaling box bedded down ready for use; note the walls are banked up high and any protruding objects are completely covered

Your box and equipment are prepared. When do you move your mare into the foaling box?
There are no set rules, and stud farms tend to move the mare any time from three days to three weeks prior to foaling; often we do not estimate the right time. However, I recommend that you move your mare into the foaling box approximately one week beforehand, which will give her enough time to adapt to her new environment and build up the required antibodies. Better to be early in moving her than too late. Once the mare is in the foaling box, make sure that you muck out thoroughly as well as skipping out at night when you check the yard, re-making the banks as described previously.

Always brush off the mare and, of course, pick out her feet *before* putting her in at night, **thus avoiding extra dust and dirt in your clean stable**. At whatever stage of pregnancy your mare should still go out in the field during the day, circumstances permitting. If for some reason the field is not usable, walk her out in hand – **she needs exercise at this time**.

When foaling is imminent, leave the following items near the foaling box at night:

1 Foaling bucket and equipment (see p57-8)
2 Bale of clean straw (for use after foaling)
3 Wheelbarrow or skip and fork
4 Headcollar, rope and tail bandage.

Make sure that if you are not in the yard during the day, the person in charge knows where the foaling equipment is kept, in case the mare foals down during the day.

Vaccinations and worming

It is advisable to have your mare vaccinated with her annual influenza tetanus booster at least one month prior to foaling (and no later than three weeks before). Vaccine given to a mare at this time should provide enough immunity to protect the foal for the first three months of its life, provided it has suckled well on the colostrum.

It is also important to worm your mare at this particular time, even if this were to come at a date earlier than the one demanded by her regular worming programme.

Caslicks

Remember, if your mare had a Caslick (the stitching up of the vulva to prevent pneumovaginal infection or air entering the vagina) it is *important* to have her unstitched by the vet approximately two weeks prior to foaling down, or as near to the foaling time as possible. Your mare *must not* foal down before a Caslick has been unstitched (sutures removed). If this is not done, she may have great difficulty expelling the foal (see 37-8). In studs where there is 24-hour continual observation and supervision they do not cut open the area until the waters are seen to burst. For a home foaling, however, it is recommended to do this earlier, in case no-one is present when labour begins. If your mare goes into labour and no vet is instantly available it is quite simple to cut an opening yourself using a sterile sharp razor blade. It is unlikely that the mare will resist, as she will probably be numb around the area of the vulva.

SUMMARY:
FOUR WEEKS TO GO, PRE-CHECKLIST

1. Foaling box ready
2. Foaling bucket and kit ready
3. Vaccinations done
4. Worming done
5. Caslick sutures – remember to remove two weeks prior to foaling
6. CCTV, video or foaling alarm arranged, if applicable
7. Days off organised (a difficult task as one can never be sure of the actual foaling date)

Feeding from four weeks to birth

See feeding advice and the chart which includes this period on p51-3; let me emphasise again that your mare's need for a higher protein diet increases dramatically in the last three months of her pregnancy. So here we have the problem of dealing with the last four weeks. If the mare is still stabled at night at this stage you will probably find that she will go off her feed, especially the morning one. Do not worry unduly; this is nature's way of telling you that the foal has grown so large that it is taking up a large part of the abdomen, so there is not much room left for bulk. Gradually reduce the protein level and bulk so that one week before foaling, you are feeding only two-fifths of what you fed five weeks previously. Droppings should be looser than those of a normal stabled horse. This will assist the foal in passing his first droppings (meconium) after birth.

The last few days before foaling your mare should be on a laxative diet. I suggest add some bran with Epsom salts, and if you can estimate the time correctly, simply feed hay and low energy feed for the 24 hours or so before birth. Do not forget how important it is to maintain your chosen vitamin (see p53) throughout. **Mares out on spring grass are naturally on a laxative diet**, and if you bring them in at night Epsom salts are unlikely to be necessary.

Mare observation

Mare observation is of paramount importance. Providing you are present when your mare foals down, you may be able to assist should there be a problem; however, it is imperative that you are on site to call for help immediately if it is required.

The mare should be closely observed during the final stages of pregnancy. On large Thoroughbred studs there is a special area for sitting up adjacent to the foaling boxes, but the personnel on duty should be quiet, and not allow the mare to know that she is being watched – she needs peace and tranquillity at this time. Today, modern equipment such as *electronic foaling alarms, closed circuit television* and *video* make observation easier:

Foaling alarm: The transmitter and sensor are placed on the mare with the aid of a leather breastplate and a roller or surcingle, and the alarm alerts you as foaling becomes imminent, usually 30–35 minutes prior to foaling. It is activated by the change in the mare's body temperature. The receiver can be used up to approximately 500 yards (457m) away from the stable. It is also likely to be activated if the mare develops colic, or becomes hot at the height of summer.

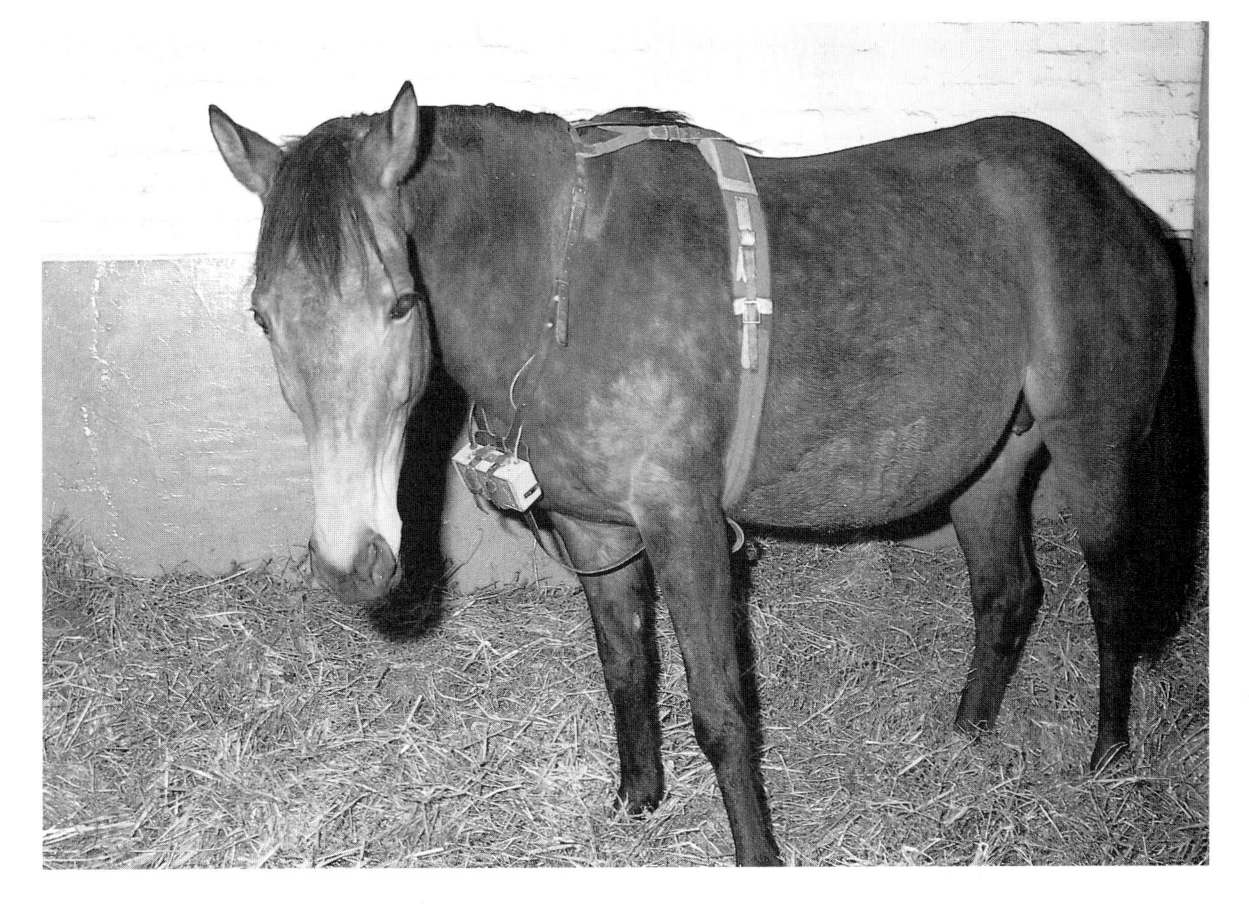

A pregnant mare with foaling alarm in position

When the alarm has been activated and labour commences, providing the mare is standing, remove the roller and breastplate – it will be more comfortable for her.

Closed circuit television (CCTV): A camera is installed in the roof of the stable and the mare can then be observed from the comfort of your home on a viewer. Today, sound is also available. By observing your mare every night you should get to know her behavioural pattern. I suggest you set the camera up at least one week prior to foaling, and during this period, note the mare's normal activity. In this way you will soon become familiar with her eat and sleep pattern. When labour is imminent, the mare's normal routine will change, and she will be seen to pace round the box and probably get up and down (see p63).

CCTV is also an ideal way for friends to observe the birth without unduly upsetting the mare. Statistically it has been proved that although the majority of mares may foal down without any assistance, the few that do have problems can, without a doubt, often be preserved from disastrous results if they have outside assistance. If you are not there to observe the event you will never know what happened. Hence the reason that most mare owners make arrangements to be on site. It has been shown that foaling normally occurs between 20.00 hours and 07.00 hours, and more often in the early hours of the morning.

Video: If you own a video camera it is now possible to purchase a booster and fix your video camera in the stable, and run a cable from there to the viewer.

How to recognise when your mare will foal down

If you are witnessing the build-up to birth, there are several tell-tale signs to look out for. However, the actual gestation period depends on the foal's growth rate. Officially the mare should go 340 days from the last date of covering (see Gestation Table p186), but nature has shown that normal births may take place anything from fifteen days early to six weeks late. A foal is only deemed premature if born earlier than 325 days, and sometimes, with good animal husbandry, earlier births may be saved (see Case Study p84). I shall now discuss the signs that point to foaling down. So much has been written on this subject, and no words will give you an accurate birth date, but you should begin close observation from about **four weeks prior to the official birth date**. Keep an eye on the following physical alterations – you should be fascinated by the various body changes:

1 The Udder: This swelling or 'bagging up' of the udder is the most obvious sign – observe carefully. The udder should begin to fill three to six weeks before birth. If the mare is stabled you will notice that the udders will be more swollen in the morning, before you put her out, than in the evening, after exercise. I suggest that you check them twice a day. Mares out at grass at this stage keep a steady udder size due to the fact that the mare is permanently exercising herself.

Most mares' udders bag up fully a few days before foaling, and the phenomenon known as waxing up should take place 6 to 48 hours before she gives birth. But beware, often the human eye does not notice this, and owners can be taken by surprise. **Waxing up** is the formation of little drops of the special early milk known as **colostrum**; the leakage from the udder coagulates on the nipple and appears like a tiny drop of wax. However, it is often difficult to identify.

If your mare actually runs milk, then you can be sure that foaling is imminent. There has recently been extensive research into the ingredients of mare's milk, especially in relation to giving birth. There are litmus papers which change colour when the colostrum is made, indicating that foaling should take place within 24 hours. 'Running milk' means that leakage from the udder is seen as the mare walks around, and evidence will be found in a dried-up form on the inside of her legs.

I feel strongly that you should *not* milk off your mare at this stage and freeze the colostrum (see *A Note about Colostrum* p79). If the mare dies at birth and you need to acquire additional colostrum urgently, you will find that most studs have some in their freezer; also there are colostrum packs for cattle on the market (see also p91).

2 The Vulva: When you begin the period of close observation, look under the tail and have a feel of the muscles and ligaments around the vulva. You will find that they will gradually relax and slacken off, so as to allow a soft, wide opening for the foal's exit when the time comes. This is also a good sign that your mare is ready to foal. When these muscles are tight she is unlikely to go into labour, although there is a marked difference between a mare who has foaled previously and a maiden mare. The lips of the vulva also soften

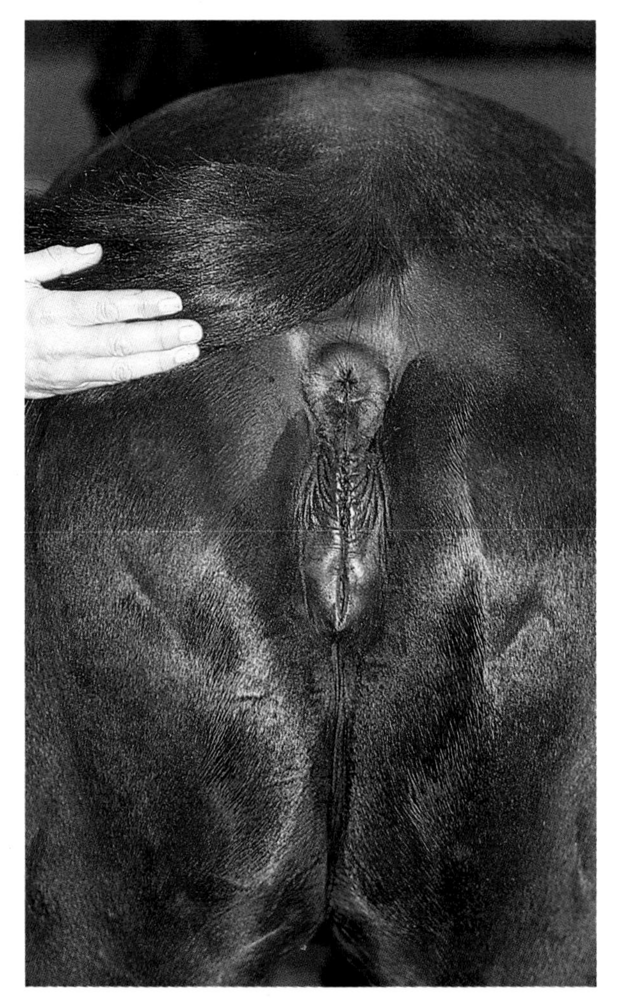

A mare near to foaling down showing hindquarter muscles slackened off and loose vulva lips

and the dividing line elongates (lengthens); they are easy to open when parted by the observer.

3 Body, Loins and Hindquarters: Stand back and look at the conformation of the loin area from the side and rear. As the weeks go by you will notice that the mare's body formation will change and 'drop'. She will get much heavier and slower. When touching the muscles over the hindquarters they will feel quite relaxed and soft. All these signs are, of course, easier to observe on a mare who has already had a couple of foals, rather than a maiden mare.

If in doubt, **call your vet**, but I can tell you now that if the mare is overdue, he is likely to tell you to be patient and wait. It is rare that a vet will induce your mare – much research has taken place on this matter and the chances of survival after inducement are less than 25 per cent (see Case Study 98).

4 Your Mare's Legs: Sometimes the mare's legs (especially the hind legs) are seen to fill a week or two before foaling. Do not worry, there is little you can do other than make sure that she does take exercise during the day. If she insists on standing at the field gate most of the day, walk her in hand – this will help. The filled legs normally go down within two or three days after having given birth.

**SUMMARY:
SIGNS TO LOOK FOR BEFORE FOALING**

1. **The udder** Observe carefully from six weeks before birth date
2. **The vulva** Look for slackening of vulval lips
3. **Body and hindquarters** Watch the change of shape, and extra heaviness
4. **Legs** May fill, and mare will become much slower in her movement

IMMINENT SIGNS

1. Udders should be full and hard
2. Waxing may occur
3. Mare may run milk
4. Unusual restlessness in the field or stable
5. Muscles and ligaments under the tail very relaxed, vulva long and easy to open when parted by the observer

SUMMARY: PREPARATIONS IMMEDIATELY BEFORE BIRTH

1. Place foaling bucket close to the foaling box
2. Check that CCTV or other observation method is in working order
3. Leave the stable light on at night to accustom the mare to light at night (required if you are to observe labour)
4. Place a bale of straw outside the stable (to add to the bed after labour); a skip may also be required
5. Make sure a headcollar and tail bandage are on hand, plus the bin liner (for the afterbirth) and baler twine (preferably not nylon) to tie up the afterbirth – see list on p58
6. Ensure that you have easy access to the feed room so that a bran mash can be made up when required

Foaling Down

Giving birth for both humans and animals necessitates an enormous effort and can be an exhausting process. Remember that your mare needs peace and quiet: it is always difficult to know what to say when friends beg to watch, and it is up to you how many people you have around – my advice is, the fewer the better, but make sure that there is at least one person close at hand to assist you.

Although labour is a gradual progression, it is always explained in three separate stages.

Stage One: Labour

When the first stage begins your mare will become very uneasy and restless. She will possibly sweat (sometimes profusely), swish her tail, kick at her belly, or will merely keep looking round pointing her nose to her stomach, indicating pain. She is likely to get up and down, and will perhaps pace round the stable. Labour has begun – the birth canal is being prepared, and at this stage the foetus is quickly rotating so that its back is uppermost and in the correct position for expulsion (see diagrams p64); it is said that this rotation of the foetus can take only 18 seconds. When your mare is quiet, **put on a headcollar and tail bandage**.

First sighting of the foetal bag

NOTE: *In order to show the different stages of labour two mares have been used for this sequence*

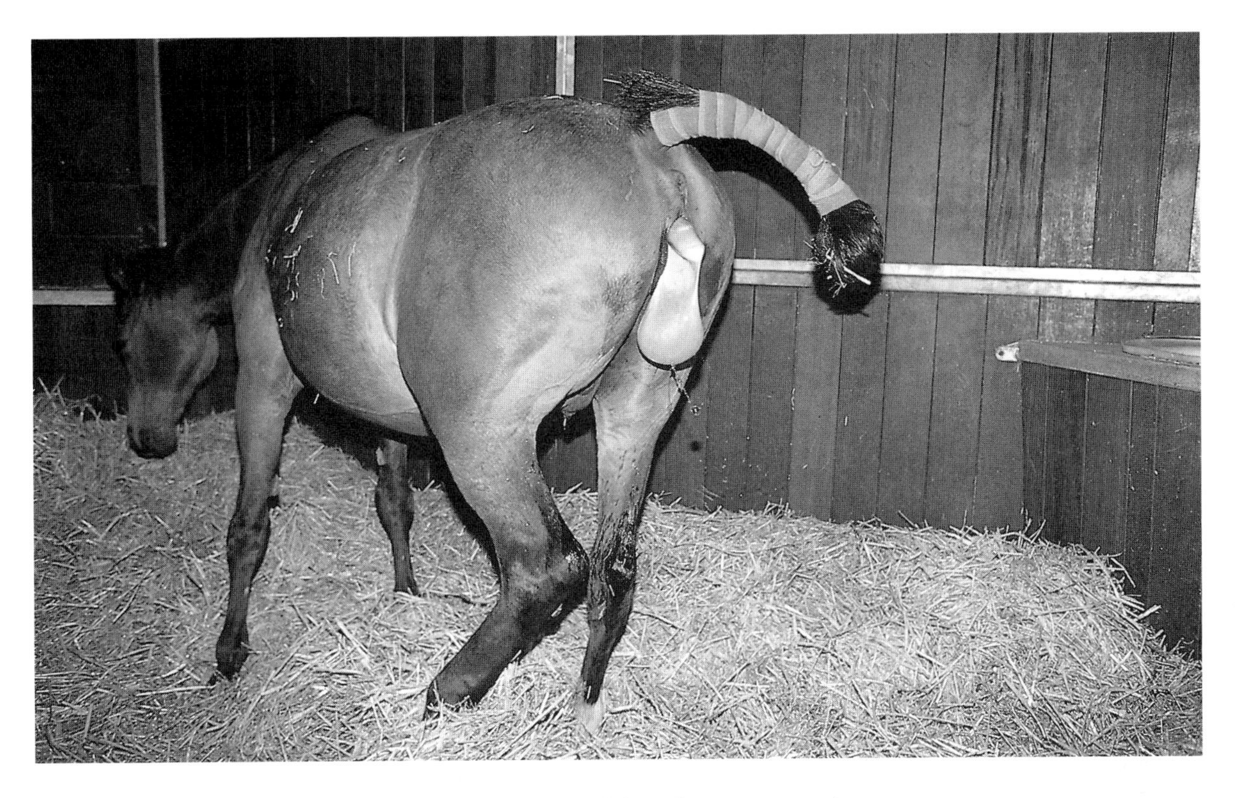

The mare will get up and down between contractions

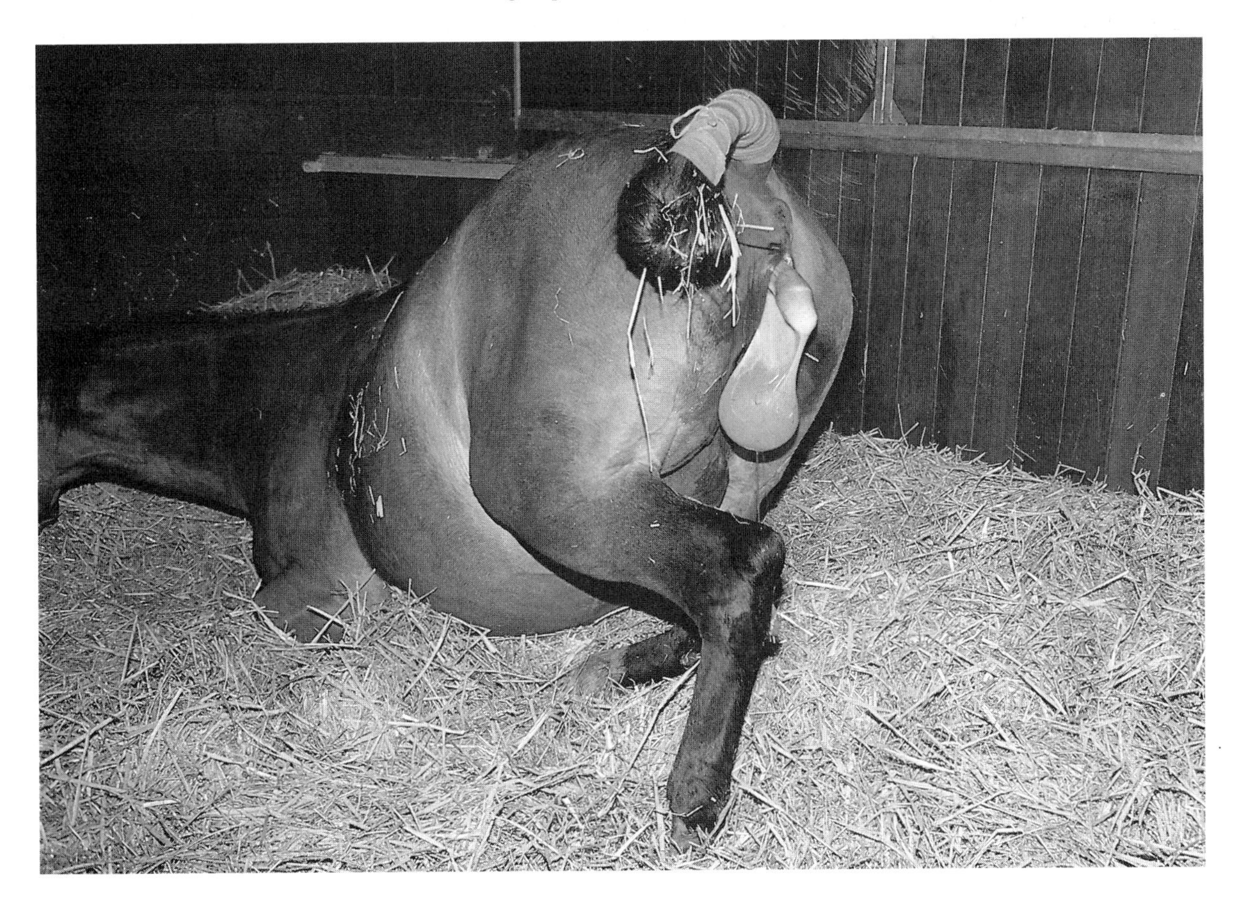

Two feet and head presented correctly. The mare now prepares for the more difficult expulsion of the shoulder

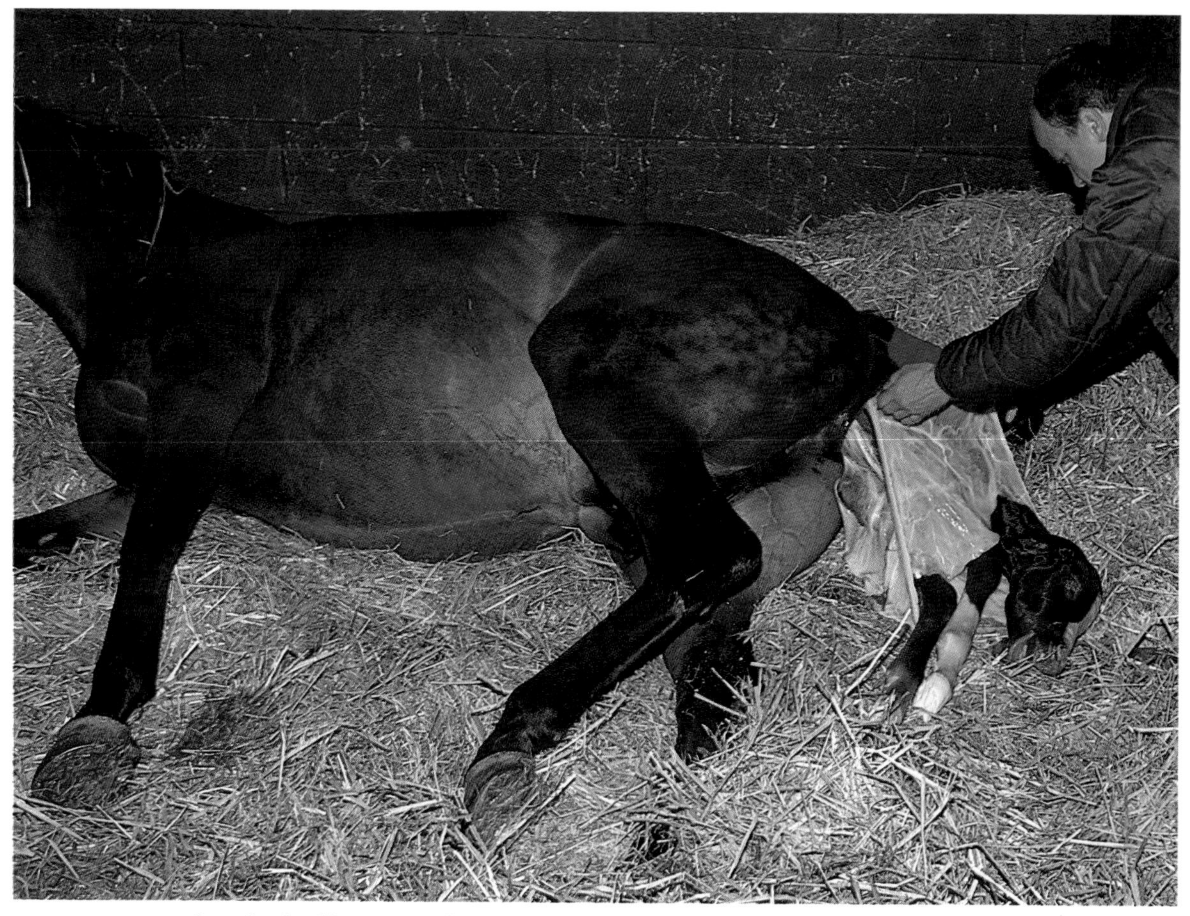

Once the shoulders are expelled you must *remove the bag away from the foal's nostrils and ensure that the airways are clear*

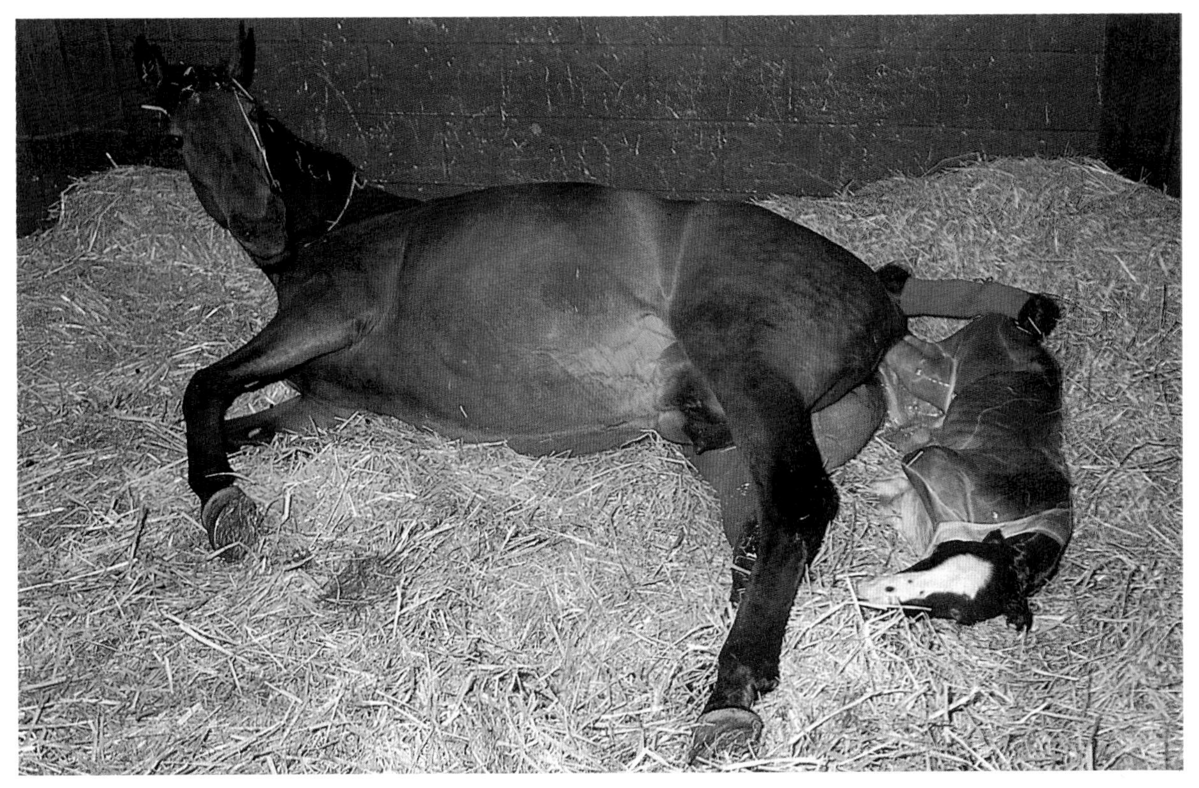

Birth is almost complete; the foal's hind legs are seen resting inside the mare

Do not disturb mare and foal at this stage; the umbilical cord is still attached

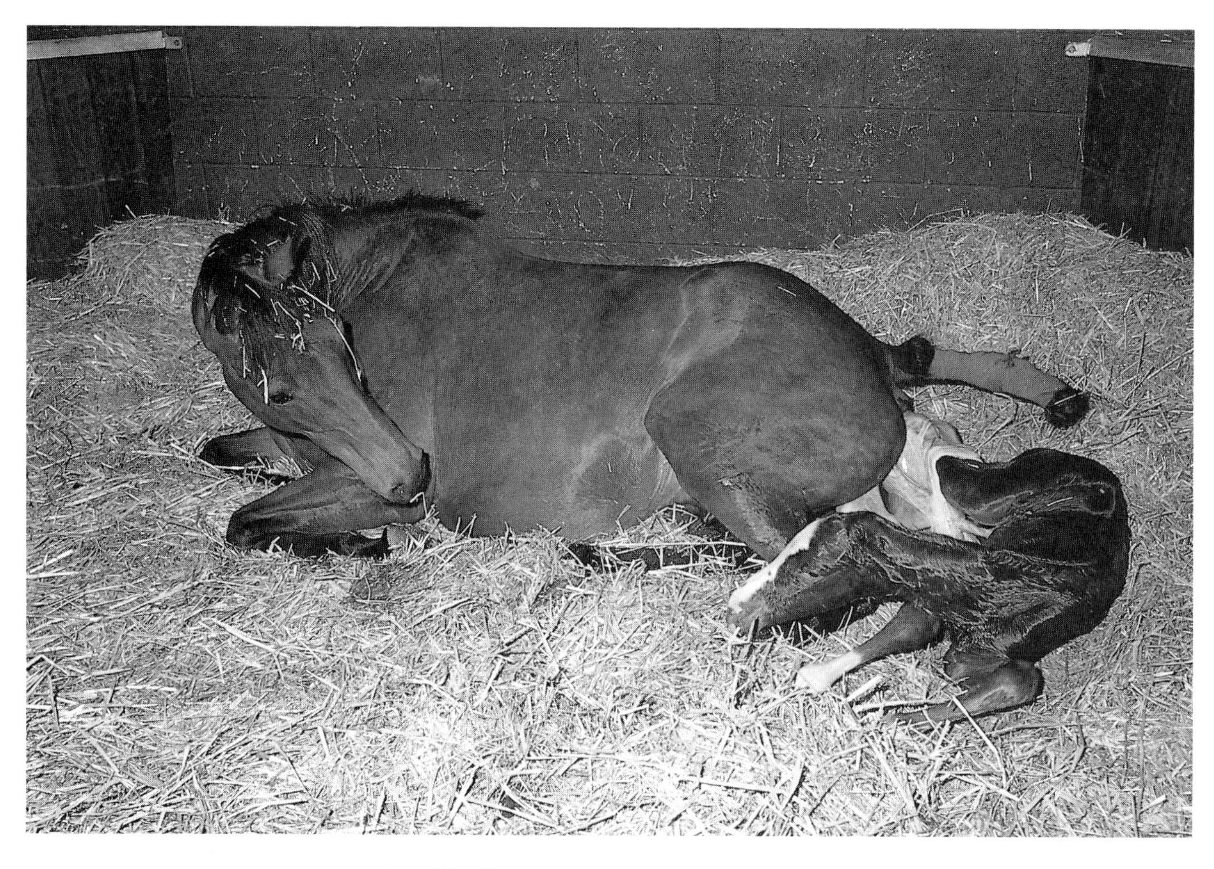

The mare greets her newborn foal

The assistant treats the umbilical stump immediately after the cord breaks

The mare meets her foal and the process of bonding begins

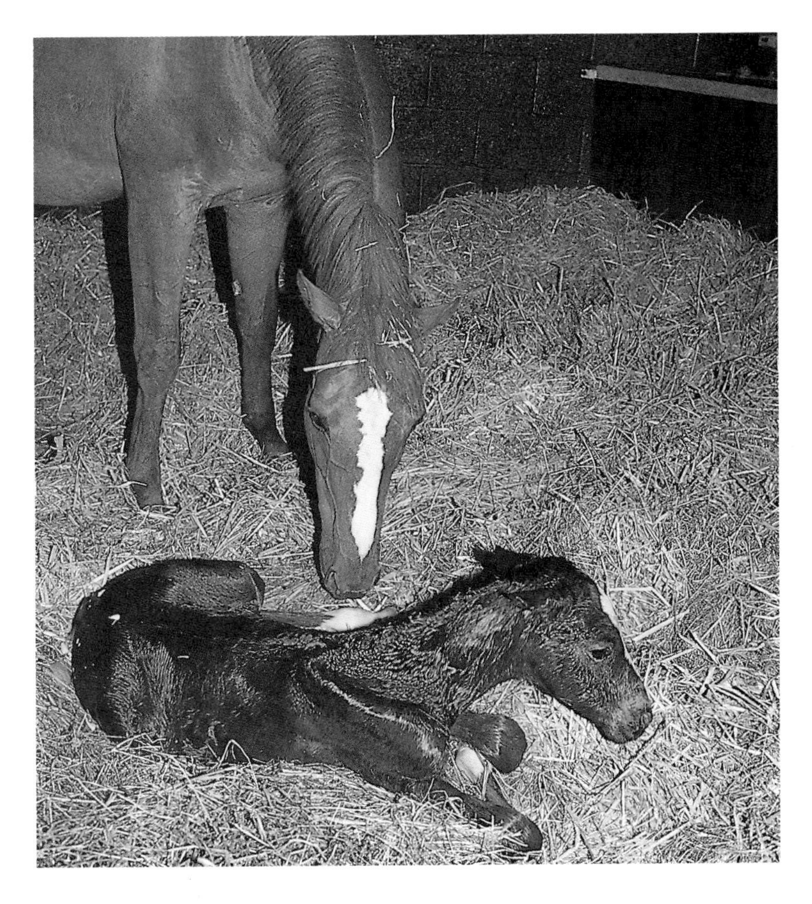

Some professional studs automatically give the foal an enema

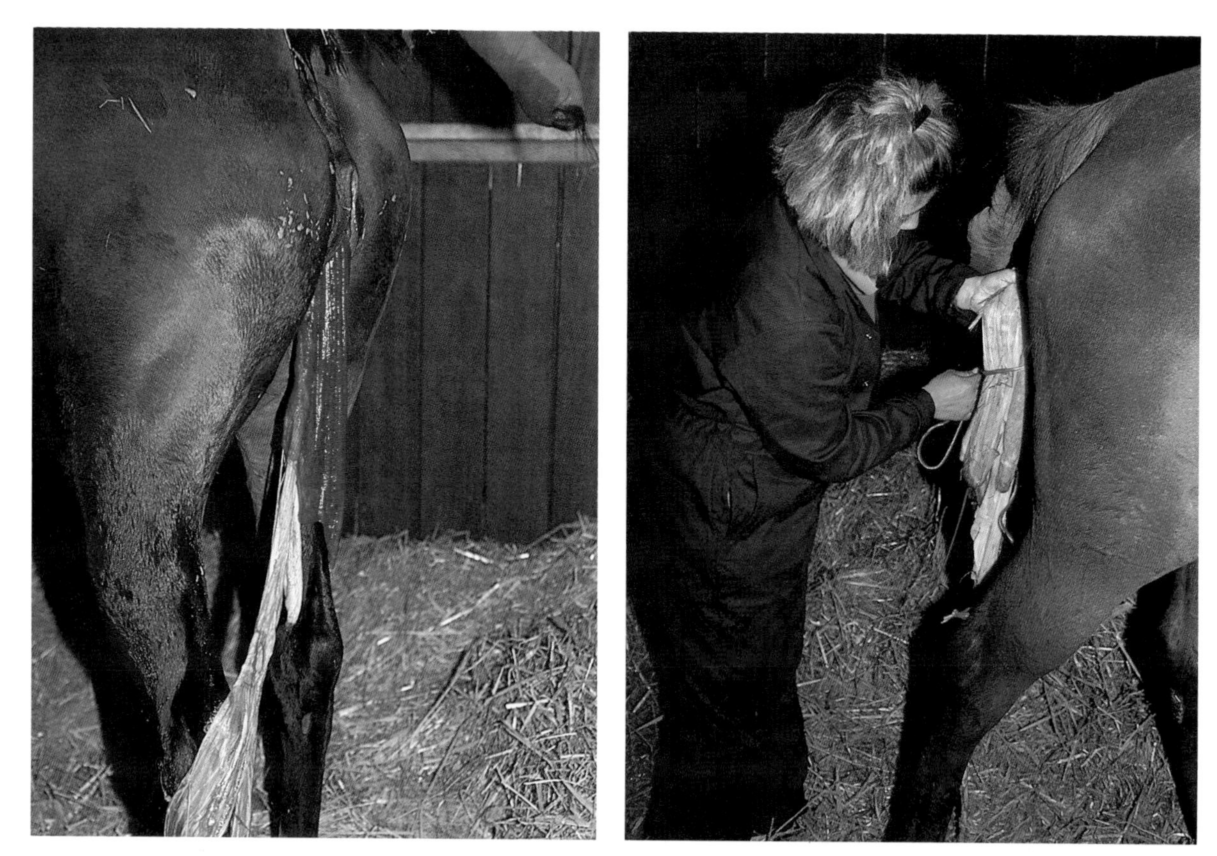

Here you can see the placenta and bag, ie the afterbirth

The afterbirth should be tied up so that the mare does not stand on it

The foal's first attempt at standing

Bonding of mare and foal

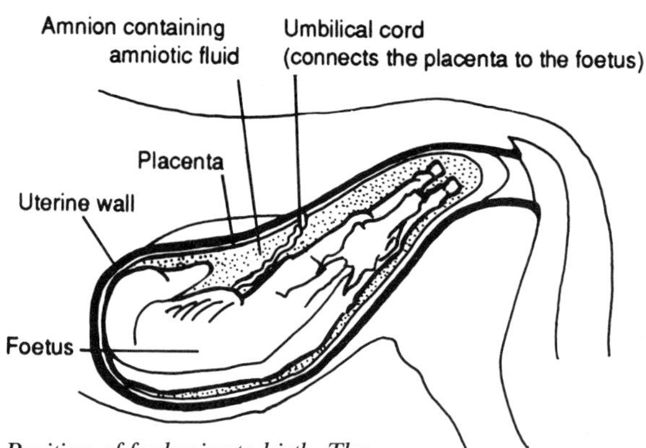

Position of foal prior to birth. The foetus gets its nourishment from the placenta which is attached to the uterine wall

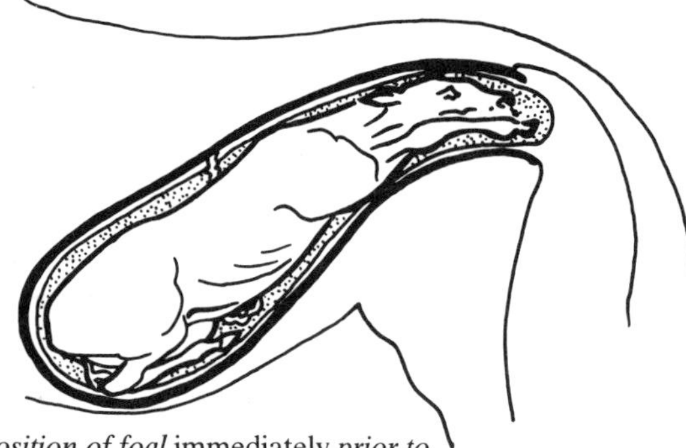

Position of foal immediately *prior to birth – note the foal has rotated to its correct position for expulsion*

Stage Two: Delivery

You are on duty. Observing from outside the stable with your foaling bucket at hand (and overalls on), you watch the waters burst. The mare will straddle her hind legs as the water membrane bursts and a rapid gush of yellowy coloured liquid is expelled (this in fact rinses the birth canal); it can be from ten to twenty litres of fluid. Be alert at this stage, but do not panic – you cannot assist yet. Keep calm and observe, labour is now taking place and it should not take more than forty minutes overall. (If 6 to 10 minutes have passed since the water bag burst and no limbs appear, yet straining continues, *do not delay*, call your vet.)

Your mare may get up and down again, probably arching her back and straining. Eventually a light-blue coloured bag will appear and, with much groaning and heaving, the first foot will show (still covered by the bag which is the amniotic membrane and provides a slippery surface to facilitate delivery through the birth canal) then, quite quickly thereafter, the other foot, and then the head, pointing forward, resting on both limbs. (You will notice that one leg is further in front of the other which slants the shoulders, hopefully assisting in their expulsion.) **This presentation is correct – if the foal is not presented in this way, do not hesitate – call your vet now**. Suspect a malpresentation if you can see the head but only one foot; or the head but no feet at all; or two feet and no head: particularly if straining continues but nothing else appears.

You should now quietly observe (from inside the stable) as the mare's contractions push out the shoulders and chest of the foal, at which time the bag should split. **Expelling the shoulders is the most difficult part** and there are **times when your assistance may be necessary**, especially if a very large foal is presented, or your mare is a maiden. It is hard to tell you when to help; all I can say is that if the mare is seen to strain continuously, and if the presentation of the foal is correct (two feet and a head), but expulsion of the foal is not progressing, this is the time that she will need assistance: take hold of the foal's legs *above* the fetlock joints and, *as she strains*, you pull. Only one person should help and *only* when the mare strains. *Never* take hold of the actual joints. Once the shoulders are out your mare will probably take a rest: **do not disturb her**.

When she is ready, further less violent contractions will take place – and your foal is born. Most likely his hind legs will still be resting inside the mare. Once again, do not disturb mare or foal at this point – the umbilical cord is still attached and the foal continues to receive oxygen from the mare's blood through the cord. If it breaks prematurely the foal could suffer a lack of oxygen, causing future problems. You may have time to notice that the colour of the cord changes from a deep purple to red, and then from red to pink. If you put your finger around the cord when it is still red you will be able to feel the arterial pulse; this will gradually weaken until it stops pulsating completely, by which time the cord will be white. *This* is the right time for the cord to break. Let nature take its course, and the cord should break naturally when the foal struggles or the mare gets up. **Do not hasten this process**. As the cord breaks the foal will take its **first breath of real air** and you must make sure that the bag is carefully pulled away from the nostrils if nature has not already done this.

Once the cord is severed, it is *vital* that the umbilical stump, which should be about 1½in (5–8cm) long, is *treated immediately*. It is the one area where infection can enter and cause problems later on, therefore an antiseptic product *must* be applied.

I suggest iodine, followed by an antibiotic powder (see p57); apply as follows: your helper holds the foal in the lying down position, and at the same time lifts the upper leg. Using surgical gloves and cotton wool, you hold the navel stump between your fingers and pour on half a small bottle of iodine; make sure that any excess is spilled on cotton wool to avoid burning of the skin. Follow up immediately with a generous amount of powder. If you do not have these products to hand, **use purple spray**, or another antiseptic. **Excess bleeding** from the break point of the umbilical cord can occur; if bleeding persists, stop it by pinching the end between finger and thumb until it stops, then apply the prescribed treatment to the end of the stump. Excess bleeding means that there is a larger-than-normal opening, and as a precautionary measure it is recommended to call your vet a few hours after birth to give antibiotic and tetanus injections (see p81).

At this point you can check whether you have a colt or filly, and look for any obvious abnormalities.

The foal should seem quite perky, even so soon after birth.

The next step is also extremely important – **bonding between mare and foal**. If the mare has not made contact with her foal, gently move him nearer to her head so that she can lick him and talk to him. I suggest towelling him dry or rubbing him down with straw to **stimulate circulation** and **prevent him catching cold.** If the mare is a maiden, it will take her a while to understand that she has just produced this 'little wonder'. **Take your time, be still in the stable**, let mother and foal get to know each other.

As soon as your mare gets up, **tie up the afterbirth** with a piece of jute string back on to itself, **not** on to the mare's tail. It should be bundled together and kept as compact as possible so that it does not hang down between the mare's hindlegs. You will need your assistant to hold the mare still. Tying up the afterbirth will prevent her treading on it and breaking it off prematurely. Particularly if she is a maiden mare, she may object to the afterbirth hanging down between her hind legs – it will feel very strange to have this wet object flapping around.

Now, three more tasks:

1 Prepare your mare a **hot bran mash** (some studs recommend adding two handfuls of glucose) – she may well eat it lying down, but do not worry, this is quite normal. Giving birth can be as traumatic as an operation, and such an occasion warrants the feeding of a bran mash. Those who would normally be reluctant to feed bran should make an exception on this occasion. The next day the mare should go on to a higher protein diet in order that she produces plenty of milk – see p83, *Feeding the mare after foaling*.

2 Cover up or remove the very wet, bloody straw, and bank up the front of your stable door so that wherever the foal struggles to get up it will not touch any concrete.

3 **Close the top stable door** to keep mare and foal warm – go away for 20 minutes, leaving mare and foal in total peace.

SUMMARY: SEQUENCE OF FOALING DOWN

1 Preparatory stage (possibly not noticed) 2–4 hours
2 Waters burst. Call vet if labour does not progress within 20 minutes
3 Amnion bag appears showing first foot
4 The head appears, and both feet can be seen
5 Shoulders are pushed through, followed eventually by hindquarters
6 Umbilical cord breaks.

The average time taken for stages 2–6 is 10 to 30 minutes. Call the vet if straining continues for more than 10 minutes with no results.

SUMMARY: SEQUENCE OF ASSISTANCE REQUIRED DURING BIRTH

1 Before or after waters burst, put on headcollar and tail bandage
2 Have foaling bucket ready, plus items required outside stable (see p57-8, 59)
3 Put on your overalls
4 Make sure nostrils are clear when forelegs and head are out
5 Assist only if necessary with expulsion of shoulders
6 *Keep still* when foal is born and the umbilical cord (known as the 'life-line') is still unbroken
7 When the umbilical cord is broken, treat stump with antiseptics
8 Towel-dry, or rub off with straw
9 Tie up afterbirth
10 Give mare bran mash
11 Put in fresh straw, making mare and newborn comfortable
12 Close top stable door and leave in peace

All this should not have taken more than 45 minutes. The actual birth, from bursting of the waters to the complete expulsion of the foal, if without complications, averages **10 to 30 minutes**. Keep an eye on your watch throughout, and if the mare is straining for a longer time without results, **call your vet** – it is better to have called him even if there is no problem, than to have put it off so that by the time he arrives it is too late.

By the time you read this you will, I hope, realise why it is so important to be prepared and to have your foaling bucket ready, plus all those other useful items outside the stable, listed on p57-9. *Everything, happens very quickly*.

If you have CCTV you can now continue to watch progress on your screen. If you do not, I would suggest you leave the light on and go back in ten minutes to look through the window – you need to make sure that the foal does not get stuck in a corner when struggling to stand, or the mare does not get colic symptoms. Note, however, that the next stage of labour, the expulsion of the afterbirth, can be similar to colic in symptoms.

Stage Three: Expulsion of the afterbirth

This may take place after about thirty minutes or up to six hours after the birth of the foal. It is expelled with a further few contractions, and the mare's behavioural pattern can be similar to a mild colic. Sometimes the expulsion is done in stages. If the mare has not passed the placenta within eight to ten hours, call your vet. He will treat the mare with antibiotic drugs in order to minimise infection; today, a vet will not remove the afterbirth manually as forced removal may cause permanent damage to the uterine lining.

When to call the vet: Here is a list of the occasions when you must call your vet. There are, no doubt, other circumstances not listed; use the following simply as a guide.

During birth:

1 When the waters break, ten to twenty minutes have passed, and nothing happens.
2 When the waters break, and you can see the head but only one foot; or the head but no feet at all; or two feet and no head: particularly if straining continues but nothing else appears – in other words, a malpresentation.
3 If at any other stage during labour the mare continues to strain but further expulsion does not take place.
4 If straining stops and the mare appears to have given up.

After birth – foal:

1 If there is excess bleeding when the umbilical cord breaks. Antibiotic and anti-tetanus injections recommended.
2 Abnormalities in foal.

After birth – mare:

1 Mare badly torn in vaginal area.
2 Colic symptoms/extended labour pain.
3 Afterbirth not expelled (after 8 hours).

Examination of the afterbirth

Once the afterbirth is expelled it is important to lay it out on a concrete area for examination. It is rather like piecing a puzzle together, but you *must* make sure that there are no small tears with pieces missing. It is expelled inside-out, and if any small part has been torn and remains inside

the uterus, it may cause infection: in these circumstances, call your vet. He will probably 'wash out' the uterus, a process of infiltrating purified water which flushes out any 'gunge' left inside. And if you do call your vet, **keep the afterbirth for him to look at**.

A minute piece of retained afterbirth can be a problem. It is known that a mare is less likely to conceive again until the problem is removed, and it may well cause discomfort resulting in mild colic. **If in doubt call your vet**. In some top studs all mares are washed out after foaling as a matter of routine, thus avoiding such problems.

Dispose of the afterbirth either by burning it, or burying it – dogs love to eat it but it makes them very sick, so be sure to dispose of it well.

(Left) *The complete afterbirth in correct form*

(Below) *The afterbirth showing a tear on its left side*

I should mention here **the Hippomane**, an oval, brown object about 6in (15cm) long, which is expelled with the afterbirth. No-one seems to know why it is always present, but the old stud grooms used to hang it outside the stable door for the first few days as the foal's good luck symbol. In Germany it is commonly known as 'foal bread', because the German scientific explanation of its presence is 'the culmination of leftover feed not utilised by the foal in the womb'.

The hippomane, seen here alongside a coin to indicate its size

Check for tears: Once the afterbirth is expelled, check under your mare's tail to make sure there is no sign of tearing. If the vulva is torn, call your vet who will come and stitch her up. Do not leave it to nature to heal naturally, especially if you want to breed from your mare in the future.

SUMMARY: ASSISTANCE/OBSERVATION REQUIRED IMMEDIATELY AFTER BIRTH

1. If you were not present at the birth, as soon as you find mare and foal, still carry out points 7 to 12 in summary on p74
2. Check foal has taken colostrum. If not, assist foal to drink (see p77)
3. Check meconium passed (see p80)
4. Check mare comfortable and await expulsion of afterbirth (see p74)
5. Check expelled afterbirth (see p74)
6. Check mare not torn (see above) or very badly bruised
7. If foal given milk from a bottle, a routine antibiotic injection is recommended
8. Check umbilical stump has dried up; if not apply more antiseptic powder

The first feed

Thirty minutes have passed – you have had your cup of coffee, but your duty is just beginning: it is your job to make sure that the foal takes its first drink from its mother's milk. You might be in luck and when you return to the stable, hopefully your newborn is up on its wobbly legs and drinking. A foal normally stands up within about twenty minutes after birth; if he cannot get up or makes no attempt to do so within about half-an-hour, there may be something wrong. If he is struggling and just cannot make it up on his own, do not hesitate, give him some assistance. Once he is up and standing you may be able to see why he has had difficulty (see example, Case Study p84).

A note about colostrum: Great importance is placed on the foal receiving colostrum, the first milk produced by a mare that has just foaled, and I would like to explain in simple terms, the reasons why. Before the foal is born he is maintained in a relatively sterile condition, then *suddenly* at birth he is subjected to *all kinds* of germs in totally different surroundings. Nature has its own way of protecting the foal against these germs by ensuring that the mare's first milk contains antibodies. These antibodies are protective substances which, when drunk by the foal, are absorbed across the small intestine (gut) into the bloodstream and have the ability to destroy germs, ie bacteria and viruses. Colostrum also acts as a laxative (see p80). The whole process ceases after 6 to 10 hours, hence the reason that the intake of sufficient colostrum has to be placed high on the list of priorities. It is known that if a foal does not receive sufficient colostrum, he may well survive, but he can be 50 per cent more prone to infection for the rest of his life.

How do you know if he has had a drink? *If* he *has* taken colostrum he will definitely show what is called the sucking syndrome, the mare's teats will be glistening, and he is likely to have milk around his muzzle.

However, I *never* rely on this – I always want to witness **this most important event** with my own eyes. If the foal goes up to the mare, and takes just two or three slurps and then goes away, this can be sufficient for the first or second feed when much effort is going into maintaining his balance; but in order to take in the right amount of colostrum you should count about **fifteen to twenty sucks by the third or fourth feed. Observe the foal's gullet and make sure that you see and hear him swallow**. And as we have seen, the mare produces colostrum for only a relatively short time, which shows how vital it is for the foal to drink soon after birth. A normal foal will be trying to drink (suckle) within forty-five minutes to an hour after birth. **So it is vital, when you return to the stable, to make sure that he has his first drink.** An experienced mare will stand still while the foal seeks the udder, and her natural instinct is to flex the opposite hind leg so that her udder is tilted towards the foal. However, some mares are so foal-proud, especially a maiden mare, that they want their foal under their nose all the time; therefore on every occasion that the foal goes to find the teats, the mother turns to face him and thus swings her hindquarters round, and he fails in his attempt to feed.

To assist the foal to drink, ask your helper to hold the mare's head while you steady the hindquarters of the foal, pointing his nose towards the udders.

Once you have made the mare stand still and the foal is looking for the mare's teats, a third person can be appointed to stand on the opposite side of the mare – as the foal tries to suck, that person should squirt a little colostrum on to the foal's nose, so giving him even more incentive to find it and take it for himself.

If your mare is not happy and is kicking out, then the person at her head should pick up a front leg to try to keep her still. **It is unwise** to twitch the mare – it could cause her physical harm. If you are genuinely stuck for help, you could tie her up, but this is not easy or recommended. If after twenty minutes you fail, let the foal rest again; do not exhaust him too much – go away for twenty to thirty minutes and only then return to try again. In cases where the mare is *very* uncooperative, you should call the vet who may give her a mild tranquilliser.

The mare who will not allow her newborn foal to drink

These mares fall into two categories: a) the rejecting mare, and b) the savaging mare.

1 The rejecting mare may well be a maiden who just does not understand what has happened to her, or who stays lying down for a longer period than normal after giving birth, or simply a nervous type who is ticklish in the region of her udders. You will have to take charge, reassure her and assist the foal to drink while your helper keeps the mare still (see above).

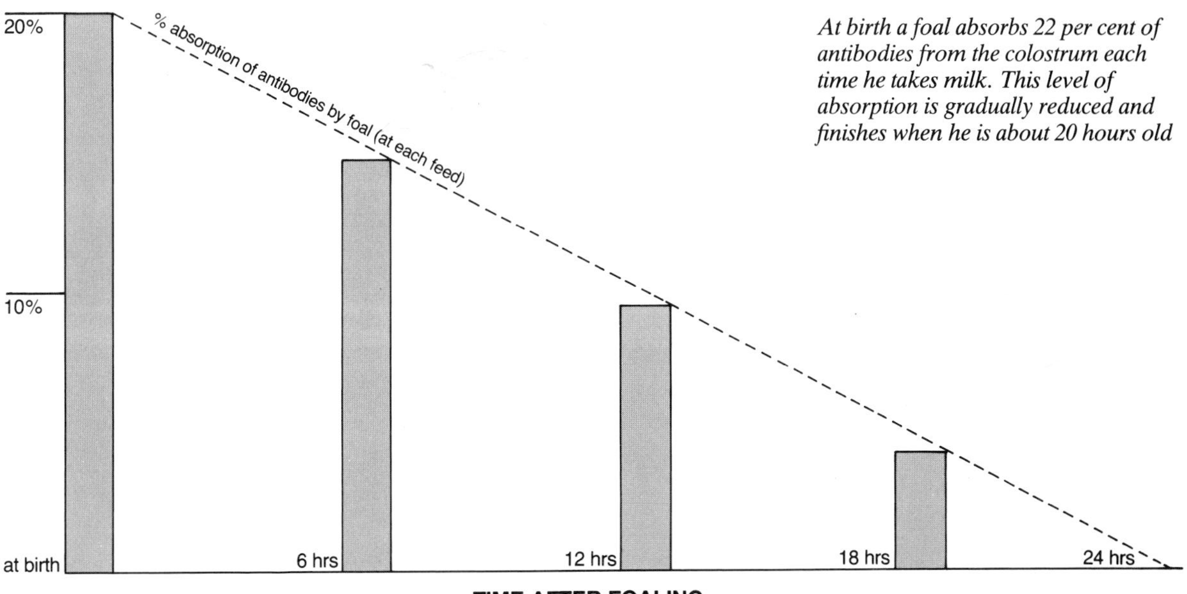

At birth a foal absorbs 22 per cent of antibodies from the colostrum each time he takes milk. This level of absorption is gradually reduced and finishes when he is about 20 hours old

Helping a newborn foal achieve the natural position for suckling

2 A savaging mare does exist, but fortunately this is not a common occurrence. A mare who kicks out at her newborn or violently threatens it requires immediate attention. She can be likened to a spoilt child and will have to be dealt with **very firmly**. The mare will need to be sedated and twitched, and then milked out by hand. Feed the colostrum to the foal by means of a bottle. Then, roughly one hour later, you should be able to sort out the problem with the help of strong assistants, and ensure that the mare will accept her newborn foal. Do not hesitate to use a stick if required.

Footnote: If you decide to breed again from the mare who behaved savagely towards her foal, it is recommended that from one month prior to foaling you touch her udders and teats daily whilst making her stand absolutely still. This may help to overcome the problem.

How to feed your foal colostrum from a bottle

One to two hours have passed. You have got a problem – the foal will not take milk from its mother. So, collect your feeding bottle, sterilised jug and a bowl of warm water and a cloth. Return after the foal has had a rest, and start again.

While the mare is held, wash off her udders with your cloth, and milk off about a quarter to half a bottle of colostrum into the jug. Then pour it into the milk bottle and, with one helper holding the mare, you hold the foal in the position it would be for suckling and have the other person around the far side of the mare with the bottle in the position of the udders.

I have had a good success rate feeding this way – after all, your foal already has the sucking reflex, it just cannot get hold of the mare's teats. Once your foal has tasted this 'magic' colostrum, you will find, if you go away again that he will rest and then be so hungry for that wonderful taste that he will suckle on his own. Providing your foal has had the equivalent of half a bottle (150ml) of colostrum, do not worry – leave him for at least two hours before checking again. By this time he should have drunk on his own. If not, repeat the same process. *If your foal has no sucking reflex, call the vet as there may be something wrong.*

Storing colostrum

Although colostrum packs are now available on the market, frozen colostrum taken from a mare within the first few hours after foaling is **the best substance to store** in case of need in the future. Once the mare's foal has taken some colostrum and you are satisfied that you have a strong healthy foal, then it will not be detrimental to this foal if you run off approximately 1 pint (0.6 litre) of colostrum to put in the freezer. It can be kept for a period of a year.

Colostrum can be taken off the mare just prior to foaling when she is waxed up; however I advise not to do so – wait until the mare has foaled down. This is because of the fact that her supply of colostrum is finite – once it has been used up

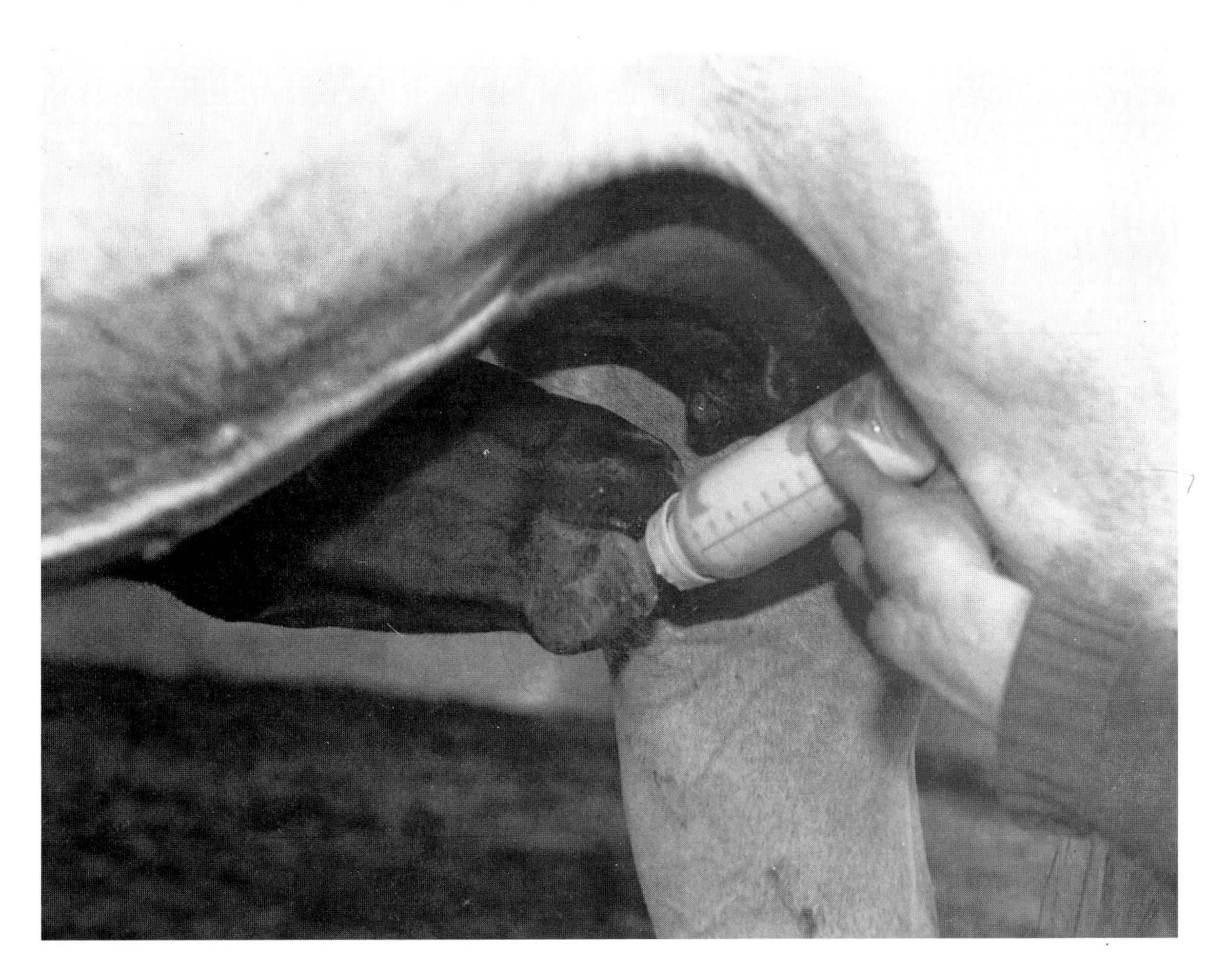

A useful way to teach a foal to take his first colostrum from the mare's udder if he finds it difficult to take hold of her teats

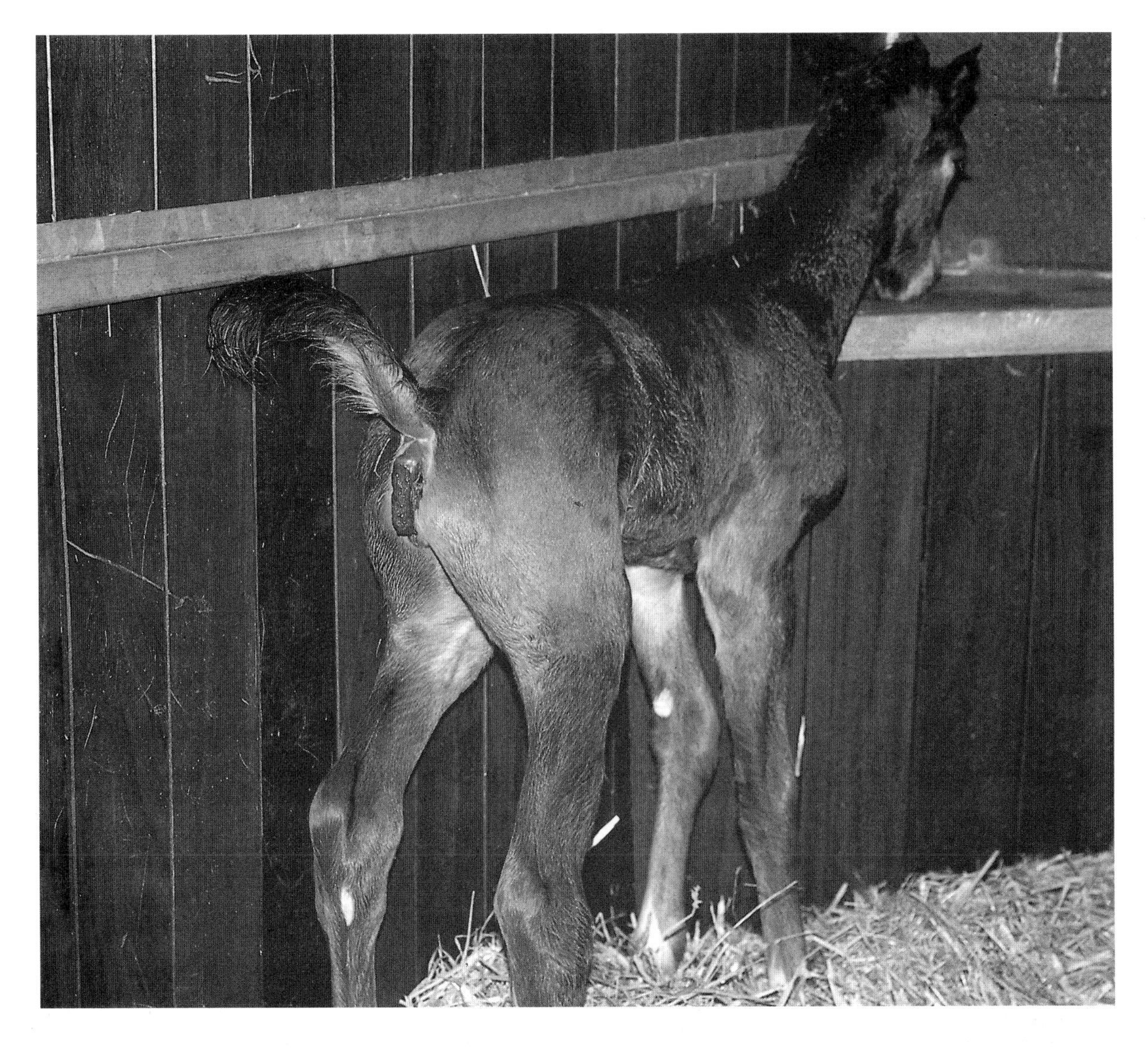

Meconium being passed

she will manufacture milk, and not more colostrum. If you 'milk' her before she has foaled you could well take too much and rob the foal.

Meconium

This is the name given to the droppings which collect in the foal's bowel while he is in the uterus. Meconium should be expelled quite soon after birth within four to six hours and certainly after the foal has taken some colostrum, which remember also acts as a laxative. To check this, lift the foal's tail – there will be some tell-tale black spots of the droppings if it has been passed. If meconium has not been passed, straining may occur, and the foal may show symptoms of severe pain, standing up, arching his back, or lying down with his front legs crossed over its head – in which case, **take immediate action**.

First, put on a surgical glove and some KY Gel or similar on to one of your fingers. Then have someone hold the foal with his tail held high. He may struggle quite hard but hold him still. Gently insert the lubricated finger into his anus and you should feel some tiny, hard droppings lodged just inside. Ease these out and straining should cease. If your mare had good exercise and a laxative diet prior to foaling, retained meconium is unlikely, but if you cannot feel any small balls and straining continues, **do not hesitate to call the vet**. He may also recommend a small dose of liquid paraffin given orally, but check with him first.

Urinating: You should note when you see the foal first urinate; the urine should be a clear colour. Once again, if there are signs of undue straining call the vet.

Post-birth antibiotic and anti-tetanus injections

At many studs it is standard policy to give *every* foal an antibiotic and anti-tetanus injection after birth as extra protection against infection. This should be carried out within 24 to 48 hours. If the birth is normal, the umbilical cord dries up quickly, and if the foal's intake of colostrum is good, then it is not absolutely vital to call a vet out to give antibiotics.

Ensure that your vet is called to give an antibiotic and anti-tetanus if there are any abnormal circumstances relating to the birth, eg foaling down takes place in a muddy field during a rain storm, or in an unprepared stable that has not been disinfected, or the mare foaled standing up and the umbilical cord broke prematurely.

Be safe rather than sorry – it is not worth saving vet's fees in these instances.

Call the vet if you experience the following problems in the foal:

1 Meconium retained
2 Diarrhoea
3 Straining
4 If foal does not stand up within 45 minutes after birth
5 Abnormal foaling circumstances (eg foaling down takes place in a field on a cold wet day). Antibiotic and anti-tetanus injections are recommended.

Remember, nature's ways are miraculous – a healthy foal will gain co-ordination very rapidly. You will be amazed how soon this leggy individual learns to find his mother's milk and scamper around the stable. Once you are sure all is in order, leave nature to take its course. **Observe but do not interfere**. Make sure he drinks at least every two hours. Count the sucks and make sure he is genuinely taking the milk down his gullet. At last you can celebrate the birth with your bottle of champagne!

A Welsh pony mare which foaled in the field with no supervision

A newborn foal looking for his mare's udders. Note the afterbirth trailing on the grass

Mares at grass

If your mare has foaled out in the field, do not worry – after all, that is nature's way. However, I feel strongly that you should carry out all the points in the summary on p76. If the weather is bad, then you must bring in your mare and newborn, or at least put them in a field shelter with some fresh straw. If the foal has not taken colostrum and you have problems coaxing him to drink, then it will be easier to deal with the problem in the confined space of a stable. Of course, if the foal has been born some hours before and has not taken any colostrum, he may be quite weak; if you are not strong enough to carry him in, take a wheelbarrow out to the field as transport.

By this time the mare will be very foal-proud, so allow her to keep the foal right under her nose at all times. If there is no sign of the afterbirth, then it may have been eaten by a fox; this is quite a common occurrence. The important thing is to keep the newborn out of bad weather and make sure he has had colostrum.

The native pony mare is much hardier than today's Thoroughbred and Warmblood, and it is highly likely that 90 per cent of such pony foalings will be without complications and will not need the kind of attention that has been described.

Care of the mare after foaling

We have talked a lot about the foal, but the mother should not be forgotten. Make sure she seems comfortable. If there is coagulated blood on her tail or down her legs, wash her off well with mild shampoo and warm water. If she squeals when the foal first sucks, do not worry – she is probably a little sore. Your observation days are *not* over; just because the foal is born does not mean you can leave all to nature. Keep an eye on your mare; it is bound to take a day or two for her abdomen to regain normality after its trauma, but if mild colicky symptoms persist it may well be that, even if you checked over the afterbirth, there could be a fragment retained inside which is causing her discomfort (see p75). A small amount of vaginal discharge is normal, but if the discharge is yellow and thick and continues to pour out as the mare walks around, infection could be present – your vet should be consulted. It is better to have her washed out now, rather than later, especially if you intend to put her back in foal.

Your mare needs exercise – a short spell out at grass or in an indoor school the day after she has given birth is advised, especially if it took a few hours for the afterbirth to be expelled.

Feeding the mare after foaling: Your mare has had her bran mash, a night has passed; now do not delay **in putting her back on a high protein diet; she needs this in order to make milk (see Feed Chart below)**. The nutrient requirements of a lactating mare increase dramatically and can be equalled to that required by a fit event horse. The nutrients are transformed into milk, of which the mare may yield up to about 10 pints (5.6 litres) per day during the first month after having given birth; this is reduced after about 3 months. Spring grass is, of course, high in protein, and is the best ingredient for milk production. If for some reason your mare cannot go out to grass (*eg* sick foal at foot) it would certainly help to cut her some fresh grass from the paddock and feed at least twice a day. Most mares will be put immediately on to three feeds per day, and in some cases may need to be fed four feeds per day in order to ensure that the mare's intake of protein is sufficient to produce enough milk for the new-born foal.

TYPICAL EXAMPLE OF A STUD FARM FEED CHART*

(Continuation of suggested diet prior to birth, see p53.) Assume feed **stud diet**.

Brood mares	Pony	Lightweight cob	Horse	Large horse	Vitamins
am	1 lb stud diet	3lb stud diet Sugar beet	4lb stud diet Sugar beet	4lb stud diet Sugar beet	
mid-day		1 lb stud diet	2lb stud diet Sugar beet	4lb stud diet Sugar beet	
pm	1 lb stud diet	3lb stud diet Sugar beet	4lb stud diet Sugar beet	4lb stud diet Sugar beet	Yes

*Chart noted three days after the mare has given birth

Case Study: 1

Winston at two weeks old, supported by a handler to prevent him walking on his fetlocks

As a 7-year-old maiden mare, Josie returned home from stud in July, confirmed in foal. However in September she had a manual pregnancy diagnosis and a blood test, and both proved negative. She was therefore put out on loan and competed in local hunter trials; she was then hunted through the winter. No-one took particular notice of her abnormal shape, and her somewhat rounded belly was put down to her being simply fat and rather unfit; her grumpy attitude, ears often flat back, and her dislike of being girthed up were apparently her not-so-pleasant nature. Eventually, however, in May, someone at last suspected that the rounded stomach might indicate something more than just fatness, so she was looked at by a vet – the diagnosis was, of course, a ten-month pregnant mare. She was immediately taken off work and returned to the stud where she was originally covered; but nothing could make up for the inadequate intake of protein over the winter months so vital for the foal's growth, or the total lack of peaceful preparation. As a result the mare demonstrated virtually no obvious signs prior to foaling down, the muscles in the hindquarters never really relaxed, and her milk bag was small.

Surprisingly, the actual birth for the maiden mare was on time and relatively easy: a small but alert foal was born and at first all appeared quite normal. However despite desperate struggles, after about fifteen minutes the foal (Winston) still could not manage to stand up. Help was given, and it became obvious immediately why this was the case – his hind legs were right down on the fetlock joints (see Hyperextension of limb bones, p109). This also meant that he found it very difficult to suck from the mare. In order to ensure that he received the necessary colostrum, he was fed from a bottle and he rapidly gained the strength to stand on his own and drink normally (approximately six hours). Further action taken:

1 The vet was called, and antibiotics were given to the foal in case of other unforeseen infections.
2 The straw bed was changed to shavings to give the foal's hind legs an easier surface to walk on.
3 The foal was carried to the field for his first few outings and Josie was kept on the lunge to prevent her rushing about and to avoid unnecessary exertion by the foal.
4 The foal's pasterns and heels were bandaged up every day with gamgee and tape for protection.

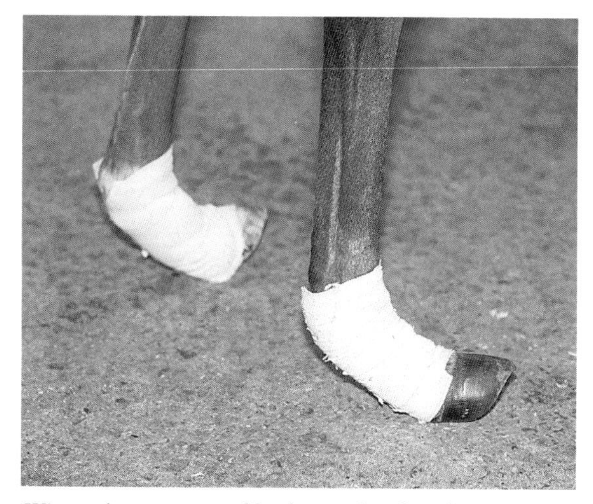

Winston's pasterns and heels were bandaged up every day for protection

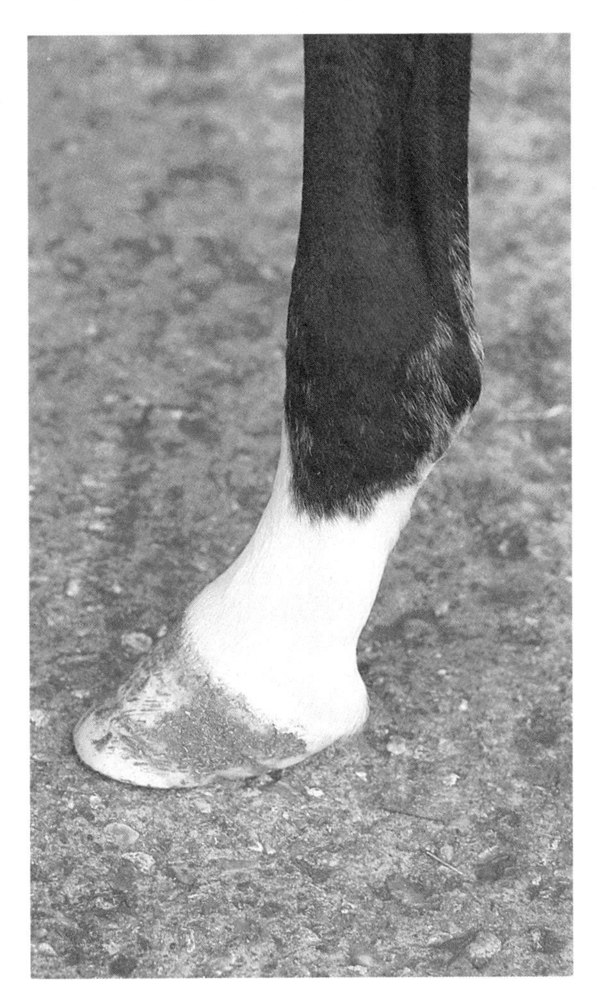

5 At two weeks old, special surgical plastic shoes were put on to help strengthen the fetlock joint and pasterns.
6 Throughout this period, and for some weeks following, a different feed chart was drawn up: four feeds a day were given to Josie so that she would produce plenty of milk and slowly regain her own loss of condition.

We will never know for sure that the reason for the foal's malformed hind legs was mainly due to the mare's fitness, but it is most likely that her strongly formed muscle structure did not allow room for the foal's growth. An expensive exercise – but in the wild the foal would have died, and even in domesticity, if supervision had not been available immediately after birth, he would not have lived long.

One hear later Josie's foal had grown into a fine, strong yearling gelding.

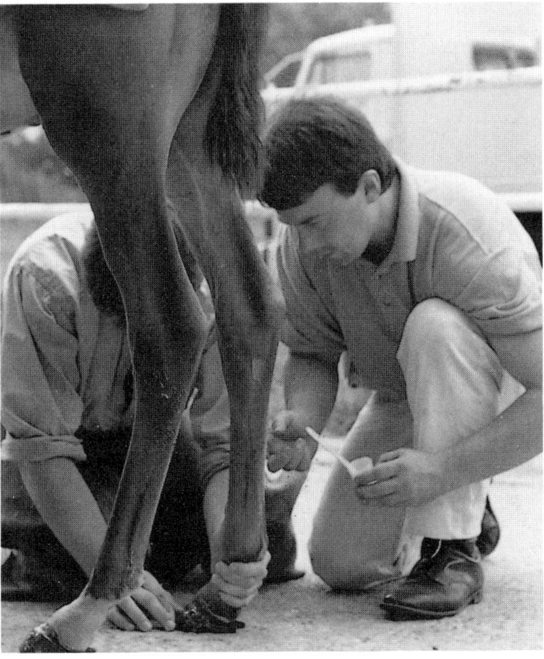

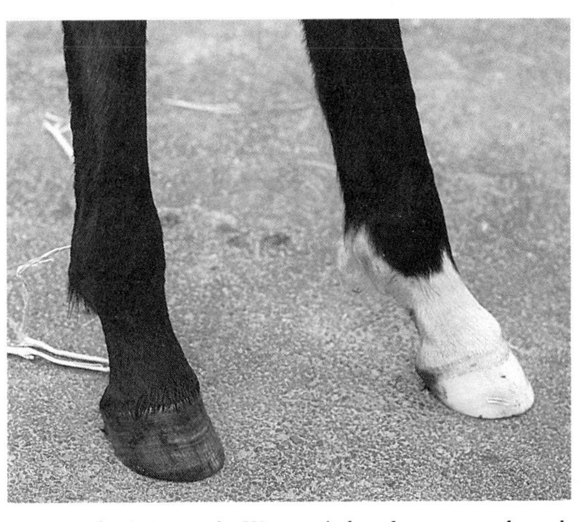

(Top left) *At 5 months Winston's legs have strengthened, although the surgical shoes have resulted in a slightly mis-shapen foot;* (Left) *The blacksmith putting on surgical shoes;* (Above) *Winston's hind legs as a yearling;* (Below) *A foal's surgical shoe*

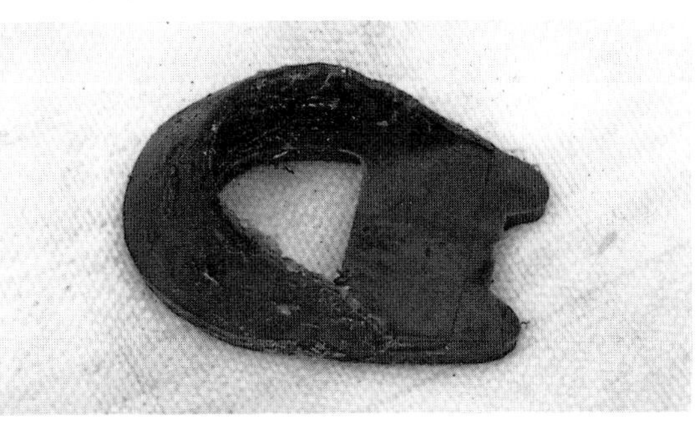

The importance of balanced additional feed, and proper care throughout the mare's gestation period cannot be overstressed: ignorant management, insufficient feeding and lack of concern can jeopardise the life of an unborn foal.

Hanna, a 13-year-old 16.3hh mare, was owned by someone who did not want her put back in foal after she had foaled down; however, due to a misunderstanding at the stud where she was kept, the mare was in fact covered again in April. Hanna left the stud in August; she was then put out to grass and basically forgotten. The grazing was shared with twenty polo ponies and so was inadequate, and hay was not fed until November. The owner never checked that Hanna was well, merely paid the livery bills; the foal at foot was not weaned, no vaccinations were given, no wormers administered, no blacksmith attended. The owner assumed all was well. However, with insufficient grazing and without the extra feed so vital to a good supply of milk, Hanna's supply soon dried up; bullied by the other ponies, her foal at foot had no chance of thriving. Weak and vulnerable to infection, when the weather became wet and cold he developed pneumonia combined with a chronic heart murmur, and died.

Hanna was 'rescued' and sent to a stud, again at livery. She was in extremely poor condition; her coat looked terrible, a sure sign of poor health and malnutrition – the underlying 'down' coat was thin, and there was no thick, warm top coat, just an overlay of 'guard hair' (long, rather wiry, rough single hairs). Her ribs were showing and her eyes were sore. But despite all this, she was obviously still in foal – the dropped belly, typical of a brood mare, was actually made to look more prominent because of her poor condition.

Hanna began her slow road to recovery in February, and great care and common sense on the stud's part was called for. She was kept in a

Hanna eight months pregnant, in extremely poor condition for an in-foal mare

Hanna full recovered with her three-month-old foal at foot

large, warm, comfortable stable for a few days; the vet inspected her and advised administering five doses of worming paste, one a day for five days, her teeth were rasped, her feet trimmed, and she was given an ad-lib supply of hay and a small feed of stud diet and sugar beet at first twice a day, then increasing to four times a day within two weeks. By this time she was ten months pregnant, and until now her intake of protein had presumably been nil. This made things very difficult for the stud manager: common sense said put in every ounce of protein possible, but at the same time the mare had to be observed carefully for fear that such relatively high feeding induced colic symptoms, swelling of legs or similar problems.

By April Hanna had begun to look happier, and the spring grass was coming through rapidly. This, combined with the additional stud diet, certainly helped her to gain some weight, although it seemed strange that she reached her full term of pregnancy – 340 days – with no sign at all of 'bagging up'. However, Nature always seems to have its way, and the foal obviously needed more time to achieve its maximum growth and development before expulsion. Thus only at 355 days (approximately one month later than normal) did Hanna's udder at last begin to swell; after a further two weeks, she gave birth to a normal filly foal – a gestation period just over twelve months. Being an experienced brood mare she knew what to do, and she had plenty of milk; in beautiful sunshine the very next day mare and foal were proudly led out to grass for an hour. Hanna was a little unlevel in front but nothing extreme. On the second day, however, when she was taken out she was very stiff; she actually appeared lame on all four legs and there was some heat in her feet, and when her temperature was taken it was well above normal. The vet was called immediately. He explained that she had given everything to the foal and left nothing for her own body and was suffering from **acute post-natal laminitis syndrome:** this is when the effort of giving birth causes the outflow of venous blood to be more than the inflow of arterial blood, resulting in acute congestion particularly affecting the hoof, since this is an almost unyielding structure.

Hanna's milk dried up within a further twelve

Hanna in good condition

hours, so the foal, in danger of starving, was put on to mare's milk replacer straightaway, first from a bottle and subsequently from a bucket (see p95). Luckily the critical 48-hour period after birth was over so the foal had taken sufficient colostrum. Hanna was very uncomfortable for a few days, but as a result of swift action and good veterinary advice, she did not die; moreover after about four weeks, her milk production had returned to near normal. Fortunately the foal never lost her desire to drink from her own mother, so Nature had its way, and as Hanna improved in health so her milk seemed to increase in quantity. There is no accurate method of measuring how much milk a mare produces, but if the foal seems satisfied then presumably the mare is producing enough – since Hanna's foal was no longer interested in the mare's milk replacer, this was evidently the case with Hanna.

A happy ending for both mare and foal, but an expensive exercise for the owner. And if the mare had not been neglected during the critical gestation period, none of this would have happened.

Conclusion:

1 Always check your mare if she is at livery.
2 In winter, never leave a brood mare without additional fodder (such as hay or dry silage), balanced feed if necessary, together with an appropriate vitamin supplement.
3 Call a vet *immediately* if mare or foal shows any unusual symptoms.

5

THE ORPHAN FOAL AND FOSTERING

All too often a mare or a foal is lost simply because its owner did not recognise that something was wrong. No-one wants to be left with a mare in milk without her foal, or an orphan, but once you can **recognise the warning signs** of illness or complications you will know when to call your vet. I cannot stress how important it is **to call the vet in time** if you need one. To bring up an orphan foal requires dedication – yet those of us who decide to breed must take on that responsibility; I hope the following guidelines on foster mares and orphan foals will be helpful.

THE ORPHAN FOAL

At birth

All breeders dread the thought of losing a brood mare, but if it does happen **do not panic**. Instead it is your duty **to take immediate action**. If the foal has lost its dam at birth, follow these simple guidelines:

1 Carry out the normal procedures given to foals after birth (see p74).
2 Prepare to feed your foal with **colostrum** – the first milk produced by the mare which contains the required antibodies to protect the foal (see p76). Collect your jug, bottle and a cloth (see p77) and **milk off the dead mare**; save all the colostrum you possibly can. (See p91 for alternative supplies of colostrum, if required.)
3 Keep the mare warm by laying a rug over her; this will prolong the period before rigor mortis sets in, and by doing this she may go on producing milk for some four to five hours: feeding the foal with its mother's own colostrum is by far the best option.
4 Keep the foal warm. If the weather is cold, cover him with a wool blanket.
5 Now feed the foal with colostrum from the **sterilised** bottle (see instructions below) – *never* feed milk or water before colostrum. If the vet is available, use a stomach tube to drench the foal with 1 pint of colostrum; repeat in one hour. This helps to get the foal going in the early stages. Then encourage it to use a bottle and/or bucket. 4–5 pints (2–2½ litres) of colostrum should be taken by a foal during his first 8–10 hours. It is recommended to feed 1¼ pints (0.8 litre) every 2–2½ hours for a horse foal, and ¾ pint (½ litre) for a very small pony (see chart, p93). Remember, cleanliness is of paramount importance: all possible infection must be avoided. I would advise wearing overalls.

To teach a newborn foal to suck from a bottle when there is no mare, take the top of the foal's head under your armpit and, with the same hand, tilt up his muzzle. Use the other hand to put the rubber teat of the *horizontal* bottle to the foal's mouth. It can happen that the foal has no sucking reflexes and however hard you try you cannot make the foal drink. He *must* have colostrum within the first four hours.

(It should be noted that some studs are having difficulty with heavy horse foals that do not want to take milk from an alive mare or a bottle. In these cases you could try coaxing a foal to drink from a bucket – it has been known to work, but beware of choking if he takes too much at a time.)
6 Remove the foal to a different stable.
7 Call your vet if he is not already present. He will check the foal for abnormalities and most likely give a shot of antibiotics for extra protection against infection and tetanus. In addition he may perform an IgG test, a blood test that measures the amount of immunoglobulin in the bloodstream; that is the name given to the protective substance found in colostrum which provides the required immunity against infection. It may not be possible to feed more than about 1 pint (0.5 litre) of the dead mare's colostrum, so by doing an IgG test you will know if this is sufficient. Thereafter mare's milk replacer should be given.

Mare's milk replacer with the equipment needed to make up the milk

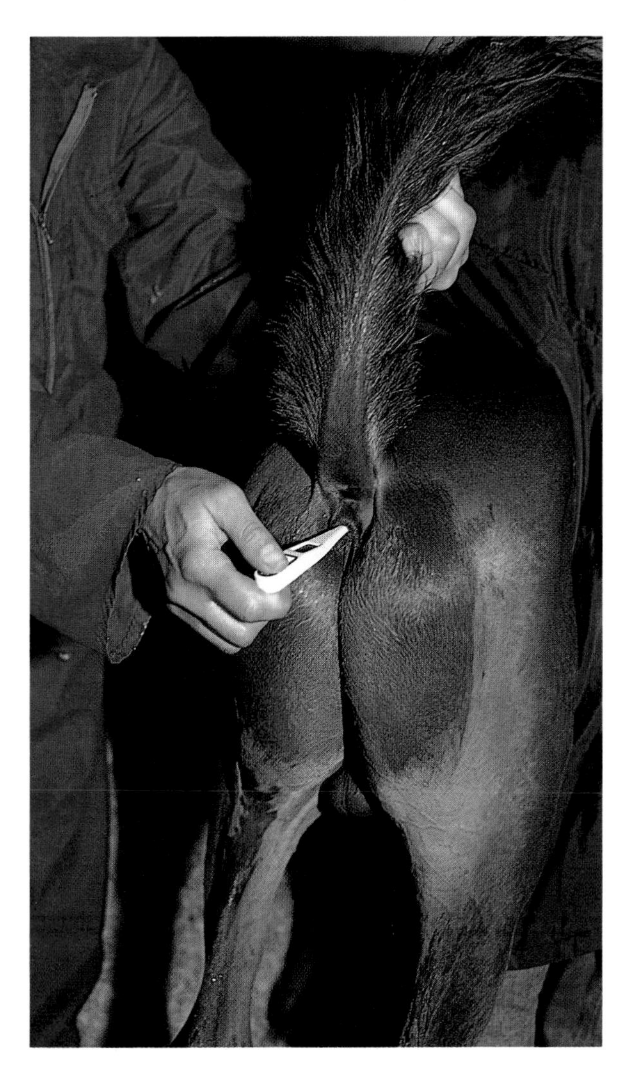

A foal's temperature being taken

8 Check the availability of mare's milk replacer. You will need to feed this within 12 hours, or earlier if little colostrum is available.

9 In the UK you can ring the organisation known as the National Foaling Bank if you need further advice; also you may wish to consider using a foster mare. Miss Johanna Vardon runs the National Foaling Bank and will organise this on your behalf, as long as a foster mare is available. This organisation operates a 24-hour service.

10 Contact the slaughterer to remove the dead mare.

11 Take the foal's temperature, pulse and respiration twice daily; anything wrong can then be spotted quickly and acted upon. The temperature of a foal is an excellent guide – if it rises ture of a foal is an excellent guide – if it rises above 38°C (101°F) or goes below 37°C (99°F) there could be a problem (see chart p107), so **call your vet**.

12 After a couple of days introduce your orphan foal to a friend, if a suitable one is available. This could be a goat, llama, small kind pony, old mare or possibly a mare with foal at foot. Sometimes such a mare will allow an orphan to drink from her udders – it certainly happens in the wild and I have seen it happen amongst groups of mares and foals.

13 You will need to feed him **every two hours for two weeks**. If you are not, for whatever reason, going to foster your foal, continue to use a bottle; feeding from a bowl or bucket at this stage may mean that he takes in too much milk at any one slurp. Try and set up a rota with staff or friends – you will not survive, otherwise! At about three weeks, reduce the feed intervals to about every three hours up until six weeks or so. To continue, see p93.

ALTERNATIVE SUPPLY OF COLOSTRUM IF DEAD MARE HAS NOT PRODUCED SUFFICIENT

1. Contact a local stud. They may well have a supply in their freezer.
2. New colostrum packs are available on the market, but seek your vet's advice before using one of these.
3. An alternative to colostrum is to feed or give blood plasma intravenously. The blood should be taken from the dead mare (or from a mare who has been kept in the same place); the blood plasma will contain the same antibodies as the colostrum. **A vet should be called in to do this.**
4. Stomach tube feed (colostrum) up to 24 hours if weak.

Mare's milk replacer: Guide to frequency of feeds and quantities required by orphan foals

It should be noted that the following chart is a guide only, and is specifically for foals orphaned at birth or in the first three weeks of their life.

Guidelines to feeding foals that lose their mares later on in their life can be found on p95.

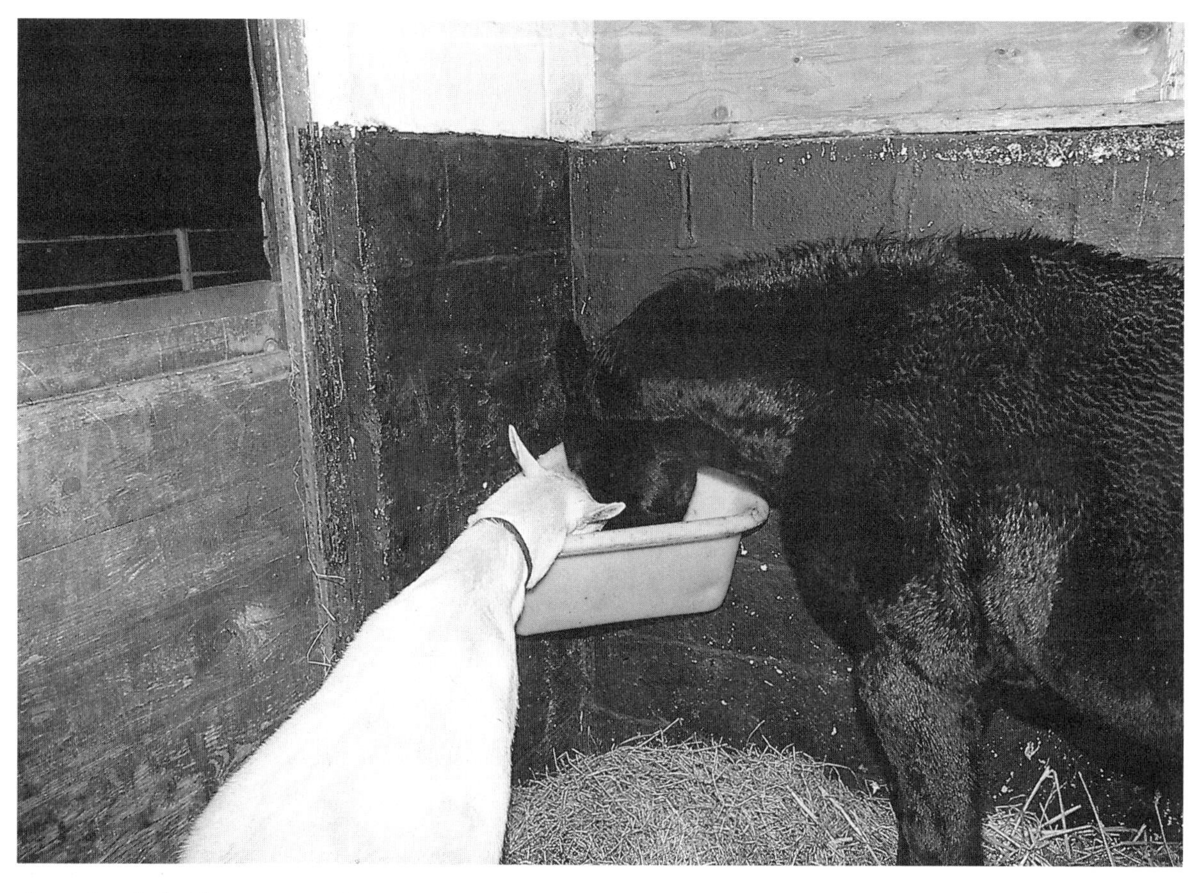

An orphan foal with goat companion

Feeding colostrum taken from the dead mare to a newborn foal

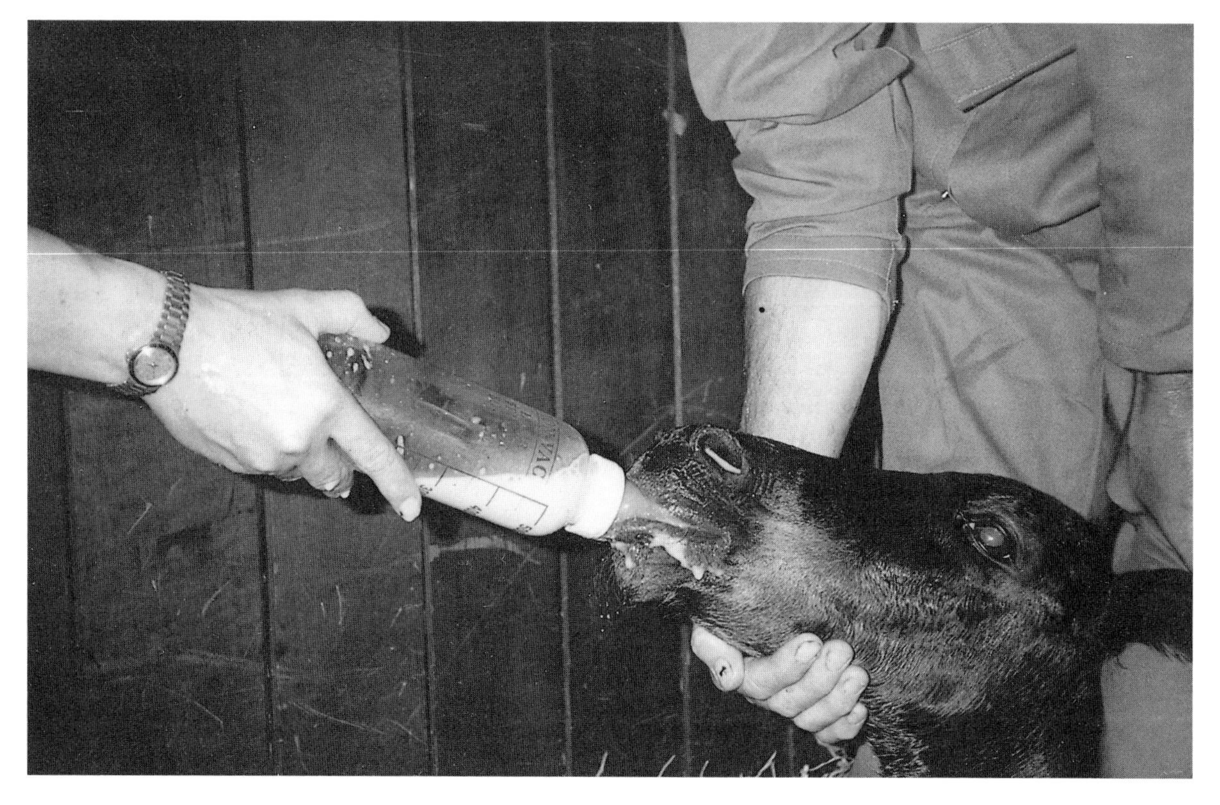

SUGGESTED FEED CHART FOR ORPHAN FOALS USING MARE'S MILK REPLACER

Age/no of feeds per day	a) Horse foal (to make 16.2hh)	b) Cob foal (to make 14.2hh)	c) Pony foal (to make 12.2hh)	Total volume per day
1st day: 24 feeds (every hour)	1 pint (0.5 l)	1 pint (0.5 l)	¾ pint (0.4 l)	a) 24 pints (13.5 l) b) 24 pints (13.5 l) c) 18 pints (10 l)
2 days – 2 weeks: 12 feeds per day (every 2 hours)	1¼ pints (0.7 l)	1⅛ pints (0.6 l)	1 pint (0.5 l)	a) 15 pints (7.5 l) b) 13½ pints (7.5 l) c) 12 pints (6.8 l)
2 – 4 weeks: 12 feeds per day but in the 4th week try to reduce the night feed to 3-hour intervals	1½ pints (0.8 l)	1⅓ pints (0.7 l)	1⅛ pints (0.6 l)	a) 18 pints (10 l) b) 16 pints (8 l) c) 15 pints (7.5 l)
4–6 weeks: 8 feeds per day, but gradually introduce a creep feed containing a milk substitute	2½ pints (1.3 l)	2¼ pints (1.2 l)	2 pints (1.11 l)	a) 20 pints (11.2 l) b) 18 pints (10 l) c) 16 pints (8 l)
6–12 weeks: 6–8 feeds per day	Increase quantity of milk by ⅓–¼ pint (197–142 ml) per feed *or* substitute with creep feed	Increase quantity of milk by ⅓–¼ pint (197–142 ml) per feed *or* substitute with creep feed	Increase quantity of milk by ⅓–¼ pint (197–142 ml) per feed *or* substitute with creep feed	
12 weeks to 5 months 3–4 feeds per day	Increase quantity according to size and condition	Increase quantity according to size and condition	Increase quantity according to size and condition	

All mare's milk replacers have been thoroughly tried and tested. Their ingredients contain the correct levels of protein, fat and sugar, plus minerals and trace elements vital for optimum growth. The milk from a cow or goat is *not* suitable for feeding to foals.

Fostering

A *foster* means '**one related by upbringing, not blood**': there are definite pros and cons to using your mare as a foster, as there are risks to placing your orphan foal with a foster mare. In my opinion the introduction of a foster mare to an orphan is a task for the professional so as to minimise any risk. An organisation such as the National Foaling Bank in the UK has years of experience and is, without doubt, successful at its job.

There are many questions to be considered regarding fostering: for example, you have found a foster mare but how do you transport your newborn? – not an easy task. If using a lorry, make up a deep bed of straw and be sure he is kept warm. One way which has been reliably recommended is to transport the foal in a feed bag in the back of a car. Also, can the foster mare return to your premises once adoption has proved successful? Does the foster mare need to go to stud? Whether to send your newborn foal away to be fostered or not is a difficult decision to make, but believe me, bottle feeding an orphan yourself is a strenuous task, with many interrupted, sleepless nights. Nature *means* a foal to have a mother, so I firmly believe one should consider adoption if at all possible. Certainly most large Thoroughbred studs put foster mares with orphan foals as the norm.

When the introduction of mare to foal is made, the personnel involved will take all necessary precautions; it is unlikely to be done without either hobbles, stocks or a specially designed partition. These will restrict the mare's movement so that she cannot be aggressive to the foal. She will be gradually introduced, and the foal's scent is of paramount importance – hence the reason for skinning her dead foal and the skin put on the live orphan. The foal will be taken to suckle every two to three hours until, gradually, the mare accepts him and eventually the two can be left alone. There are rare cases when a foster mare will accept an orphan foal without having to have her own skinned.

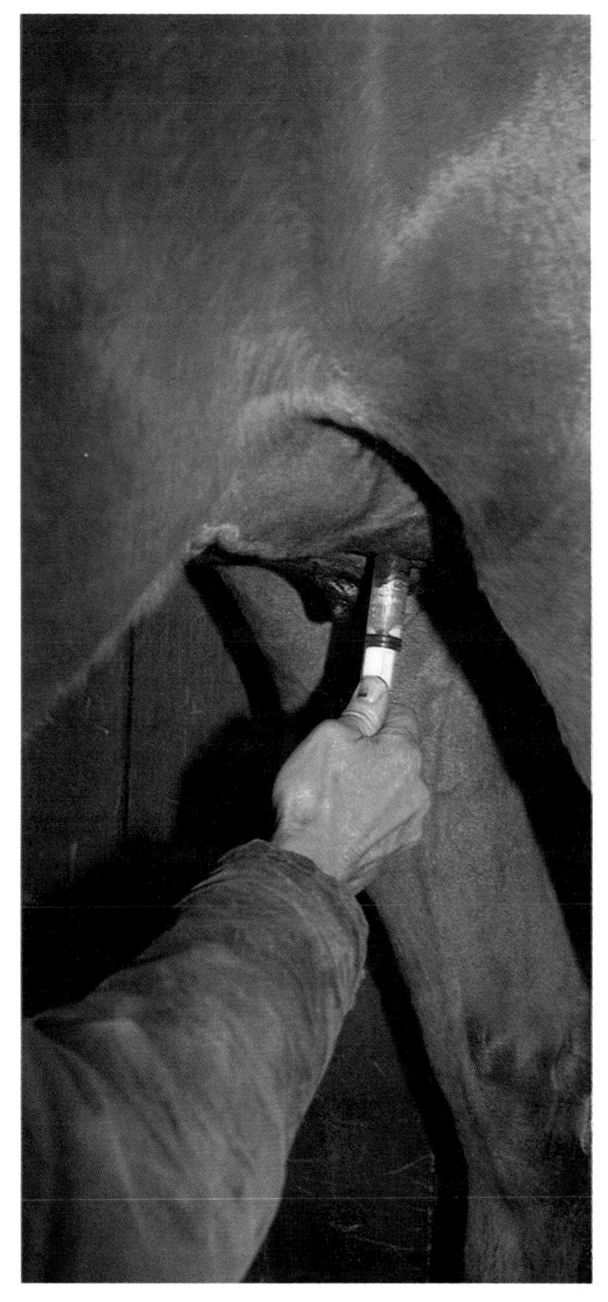

Milk can be taken from a mare using a 20cc syringe; cut off the end of the syringe before use

Your mare has lost her foal

Your mare has given birth to a dead foal – this is always very sad, but how do you look after your mare, and what do you do with the dead foal? She will be running milk – if she is to be a foster she will have to leave soon. The dead foal will have to be skinned by the slaughterer. Your instant reaction will probably be 'No, let us keep her at home', but just bear one thing in mind: if there is a suitably-aged orphan awaiting a foster mother, by allowing her to be used, you may well be doing the right thing for her **mentally**. At least she will have five months with a foal at foot after the long eleven-month gestation period.

If she is going to be sent away, you will have to sort out the following:

1 insurance
2 who pays for transport
3 whether she will be at a stud where there is a suitable stallion should you decide to breed from her again; and finally,
4 form a contract between yourself (the foster mare owner) and the orphan foal owner.

Action to take immediately after the birth of a dead foal (in order of priority)

1 Feed the mare a nice warm bran mash.
2 Milk off the colostrum and store it in your freezer in the sterilised container. It may assist another foal in need.
3 Leave her quietly in the stable with the dead foal for at least two hours so that she becomes accustomed to the fact that her newborn is not alive. Some experts advise that the mare should remain with the dead foal from 12 to 24 hours, but in my experience the mare loses interest very quickly. Use your common sense in this instance.
4 Inform your national fostering organisation (in the UK, the National Foaling Bank) and advise them that you have a possible foster mare (see p93).
5 Once you are happy that the mare is totally disinterested in the foal, move her out of the foaling box to a stable elsewhere. Close the stable door. She may be happiest out at grass for a few

hours, although do not put her out on rich grass because, unless she is going to foster a foal, you want to decrease her milk supply as quickly as possible. She will be happiest with another mare or good friend, one of course who does *not* have a foal at foot.

6 Milk off the mare approximately every 6 to 8 hours, depending on how full her udders are. If they are hard, then you definitely need to take some milk off, but if they are fairly soft and milk is running naturally as she walks around, then leave well alone. Note that the more you milk off the more she will produce, so you need to find the happy medium and give relief to any pressure, yet not encourage any increase in production. Very hard udders cause great discomfort to the mare and can result in **mastitis** (see p129).

7 Provided all was in order with the afterbirth, and the foaling down was not difficult, there is no reason why you should not send your mare away to stud after a while. However, discuss the circumstances with your vet who will advise you when, or if, she may be covered again the same season.

What to do with the foal

Your vet has probably been called, and you will probably confer as to whether a post mortem is required. He will give you the name of a local slaughterer who will collect the dead foal (be prepared for a small charge). If a post mortem is required, the vet will arrange to go over to the slaughterer's base to carry this out. If abortion is suspected, move pregnant mares to a fresh paddock and disinfect the stable area thoroughly.

If your mare is needed as a foster mare and you have made a positive decision to this effect, the dead foal will have to be skinned in order for the skin to be initially placed on to the orphan foal, thus assisting the mare to accept the orphan. Time is of the essence – **do not delay**.

The older orphan foal

If a mare dies **after the first two or three weeks of her foal's life**, fostering does become more difficult, and it may be advisable to manage the foal yourself. He should be quite tough and strong by this time, accustomed to the environment and old enough to drink from a bucket. To teach him to drink from a bucket, first use a shallow bowl and one which is *light in colour*; it is known that a foal is less worried about lowering his head down into a *light-coloured object* than he is into a dark one, so black buckets are not suitable. If he is hungry enough he will adapt to the bowl quite quickly, and then gradually learn to lower his head into the bottom of a bucket. Hold the bucket at about adult waist height – this is the easiest level for a horse foal to drink from, though obviously it must be at a lower level for a pony foal. **Do not place it on the floor**. *Remove the handle* to avoid any danger. Once he is familiar with his feed bucket you can fix it up against the wall at the same height and just tip his warm milk into the bucket at each feed. But **make sure that you throw away any unfinished milk** before the fresh feed is given, and clean out the bucket.

In order that you can have a few hours sleep, put double the amount of milk in at midnight so that he can help himself when he is hungry; he is old enough that the temperature of the milk is no longer critical. If he is in a barn with another mare and foal you will find that neither mare nor foal will touch your orphan foal's milk, although they would eat creep pellets.

Routine is important **but do not pet your foal** – he will already be treating you as a mother rather than as a human being to be respected, so make sure he is taught manners – unruly orphan foals can be quite a handful. The more he can mix with other foals, the better.

Foal creep feed can be gradually introduced at about eight weeks, or as recommended by the feed manufacturer. You should, however, continue to feed mare's milk replacer up until about twelve weeks, by which time five daily feeds should be adequate, provided the foal is out at grass most of the day.

Do not forget throughout this time to administer wormers and vaccinations as recommended for normal foals; and remember those feet.

It is of paramount importance with an orphan foal to make sure that he does not '**go into decline**' – in other words, **do not isolate him or leave him**

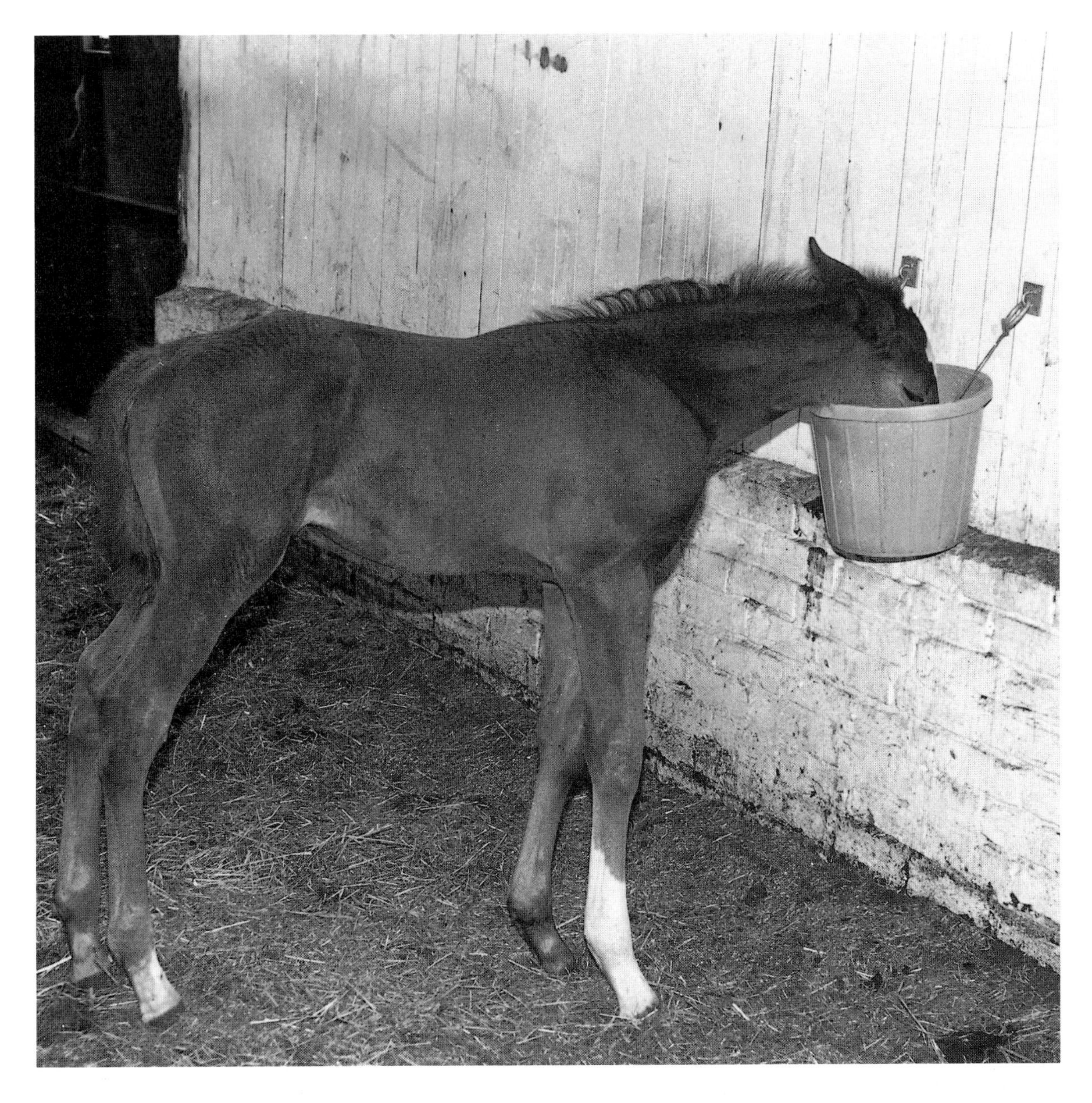

(Above) *As the orphan foal gets older he will learn to drink the mare's milk replacer, and eventually a foal creep feed, from a bucket attached to the wall*

(Left) *An orphan foal being fed mare's milk replacer from a light-coloured bucket: foals do not like putting their heads into a dark-coloured object*

on his own without company for too long a period. In the wild amongst a pack, a strong orphan will survive and there is every chance that if your foal is with a number of mares, one of them will allow him to drink from her. If so, this mare may well need extra feed as she will henceforth be feeding two foals!

Finally, I wish you lots of luck – there are many stories told of orphans that have grown into fine strong horses.

Case Study: 3

This event took place at the farm of an experienced breeder, and is a story that we might all read and learn from.

The time of year is mid-summer, and three brood mares are grazing together. One is heavily in foal, fully bagged up, and the previous evening a little wax was apparent on the end of her teats. She was stabled that night and observed every two hours. On the following day she was turned out, but at midday the owner needed to go into town, so she asked her farm worker to keep an eye on the mare. Some time later he noticed that the mare was getting up and down, and on closer examination, he saw that she was starting to foal. He panicked, called the girl groom and requested a headcollar, and led the mare to her stable. **Once disturbed**, labour apparently stopped: the mare appeared quiet and did not get down again. One hour later the owner returned and was told of the course of events. She immediately took a look and the only apparent change appeared to be a slight discharge from the vulva. However, she decided to call the vet who, on examination, confirmed that labour had begun – he could 'shake hands with the foal'. He decided that nature should be given a chance, and advised leaving the mare in peace in the hope that she would give birth. If, however, there was no change within four hours, he would return.

Unfortunately labour did not recommence, so four hours later the vet induced the mare; within fifty minutes she was foaling. He broke the waters immediately, though when the feet were presented they were moving (running). He helped the mare by pulling off the foal and once it was born it seemed quite alright, with a normal sucking reflex. However, it was not sufficiently co-ordinated to manage to suck from the mare's udder, so her colostrum was administered by bottle (see p78).

Quite soon after, the foal was able to take his mother's milk, and for the next 24 hours things went well – a great relief; until it became apparent that the foal had gone off the suck, the mare was running milk and the foal began to rush around the stable banging its head against the walls. The vet returned immediately; he introduced milk into the foal's body by means of a stomach tube, and injected multivitamins and antibiotics. Unfortunately the foal deteriorated rapidly, he had a high temperature, was shaking and appeared blind. Within hours he had died, probably due to heart failure.

Conclusion:

1 *Never disturb a mare in labour* – it is better that the foal be born outside in the mud, and then brought in immediately after.

2 The horse is naturally a herd animal; if a mare senses danger she will react accordingly. In this case the human intervened and as a result the mare's body reflexes reacted accordingly by ceasing labour.

3 *Mares need peace and tranquillity during foaling.*

6

MANAGEMENT OF MARE AND FOAL AFTER BIRTH

Once your mare has given birth to a pretty, alert, long-legged foal it is so very easy to become complacent. The long-awaited event has taken place and all seems well and looks good for the future – but how wrong you can be! The first three to four weeks in the life of a foal are a most critical period: at this time it is exceptionally vulnerable and susceptible to disease, and keen observation on your part is vital.

It is important to learn quickly how to handle your foal, to know what to look for, and how to avoid potential problems. In addition to routine handling I shall describe, in non-veterinarian terms, a number of ailments that frequently occur in the early life of a foal, followed by some of the more common limb abnormalities.

This chapter also takes a look at how to keep your foal happy and healthy during its first five to six months; what to feed; the importance of regular worming; when to vaccinate, and the farrier's visits – these are all an integral part of your foal's development.

In-hand showing is a serious commitment for many breeders. It is a great way to show off their homebred stock to the general public, and for the single mare owner it can be a very rewarding exercise. Details are discussed on how to prepare for the show ring, the type of equipment you need, and the pros and cons of taking out your mare and foal.

THE HEALTHY MARE AND FOAL

Out to grass

Mare and foal seem to be doing well, when can they have their first outing in the field? This is entirely dependent on the time of year and the weather. If it is spring or summer and the sun is shining, the day after birth is certainly in order. In fact, if birth occurs during a hot spell, mare and foal should go out a few hours later. *Do not* however, turn out if it is raining or there is a biting cold wind. Don't forget your foal is very vulnerable at this time and he **must not be subjected to rain or strong winds**. If bad weather persists, you could turn out in a safe indoor school, or alternatively walk in-hand at least twice a day.

To take your foal for its first outing in the outside world find a clean stable rubber or tea towel, fold it like a neck scarf and place it around the foal's neck. Lead the foal, as shown in the photograph, helping him to keep his balance. Remember, your mare is most likely foal-proud, so

The first outing for mare and newborn foal. Make sure that the foal is kept very close to the mare, so that she does not get upset

CHECK-LIST FOR MARES AND FOALS OUT AT GRASS

Mare

1 Check normal signs of health *eg* eyes, limbs etc.
2 a) Check udder – make sure that it looks normal. **If she is running milk** it is very likely that the foal is not well. Take his temperature and call the vet.
b) Check the udder to make sure she has **enough** milk. If she doesn't appear to have very much – and a sure sign that she hasn't is if the foal keeps tugging at her udder – then it may be that she and/or the foal need supplementary feed; ask your vet for advice.
3 Always look out for discharge from the vulva – a tiny piece of the afterbirth may have remained inside which may be the cause of the discharge. This can be quite normal for a few days after having given birth, but I suggest you ask the advice of your vet if a discharge persists.
4 Check for swelling of legs, lameness or heat in all four feet (*especially* immediately after birth). This might be acute post-foaling laminitis **and could be serious – call your vet.**

Foal

1 Check normal signs of health *eg* alertness, eyes, limbs etc.
2 If you observe straining – take action (see p57).
3 If you observe scouring – take action (see p104).
4 If you observe a lethargic foal who doesn't want to get up if lying down, don't panic immediately – if you are concerned, take his temperature, and call the vet if in doubt; though you must learn to differentiate between a foal who is merely 'sun-worshipping' and one who *is* poorly.
5 Lameness: it is unusual for a foal to be lame; he could have a bruised sole – so, observe – and call your vet if in doubt.
6 Swollen joint: **call your vet immediately.**

In all these cases, if the weather is adverse *ie* cold, hot or wet, bring mare and foal into a stable. Observe. Take his temperature. Call the vet if you are concerned.

keep the foal right by her head. It can be dangerous to let your foal run loose in the yard, but once in the field he will learn to find his own balance.

For the first outing keep the mare on a lunge line whilst she grazes and the foal will scamper around without over-exciting himself. Foals become hot very quickly and there is nothing worse than not being able to catch your mare if she goes galloping off and tires out her foal. This is especially important with maiden mares – it is all new to them. When you do let the mare loose, leave her headcollar on as you don't know how she is going to behave with her newborn; she may be difficult to catch, and if the skies suddenly open, you want her in in a hurry.

During the foal's first week, one to two hours grazing twice a day is sufficient, but if the weather is nice and warm, this can be increased quite considerably. Even in spring, after about four weeks, they should be able to stay out all day. However they will not be able to stay out at night until the nights are warm – normally mid-May.

If mare and foal **are permanently out at grass**, *never forget* to check them **on a daily basis** (or **twice a day**, if the foal is very young): things can go wrong very suddenly. One person should take the responsibility of checking them first thing every day.

For owners of mares and foals which are permanently out at grass I would recommend that they bring them in once a week for handling, including picking out their feet. If you are going to show in-hand, then more than once-a-week leading will be necessary (see In-hand Showing p115).

Teaching the young foal to lead

When the foal is about three days old, I suggest you put on a Dutch foal slip. Make sure it fits well: if it is too loose he could scratch his ear and get his foot through; if it is too tight it will make him sore. A few days later you can begin to teach him to lead from the foal slip. I recommend using an old piece of lunge rein about 3 feet long (1 metre), or a lead rope **without a clip** on the end; if you do have to let go of the foal, there is no danger of the rope being caught and causing an accident.

Your foal needs to learn to trust you. When he is five or six days old, spend some time in his stable or in the field talking to him. His fear will gradually disappear and then education will come more easily.

To teach your foal to lead is quite simple, but you do need a helper and some common sense. You have already led the foal in front of the mare using a stable rubber around his neck and you have fitted the foal slip. So the next morning use

MUSCHAMP
STUD

the piece of lunge line already discussed, holding it in your left hand, and still continue to put your right arm around his hindquarters. Gently loosen the pressure of your right arm and if he stops, just pat him gently on the area where he is used to having the pressure of your arm. On each outing to and from the field he will get better and move forward more freely; he will become braver too, more accustomed to the outside world and its strange objects. If you do not have an assistant available every day, then your aim is to teach the foal very quickly to lead beside your mare with you on the off-side and your mare being led in the usual way on the near-side.

As soon as you think he is ready, place yourself between mare and foal in the manner described and ask your helper to walk behind the foal. Each time he hesitates and drags back she should tap him on the hindquarters. He will soon understand. But do not use a whip or any other object: a kind, encouraging voice is what you need; and remember, do not hold on if he pulls back as he could go up and over, which can be harmful. Having accomplished his first important lesson you have now made life easier for yourself. If you are on your own, you can lead mare and foal side by side.

(Left) *Leading the mare and foal together*

(Below) *A foal feeder with adjustable bars that can be positioned to allow only the foal to feed from the manger; the mare cannot reach the feed*

However, if there is no-one around to help and the foal is difficult to lead, it will probably be alright if you let him run loose as he is unlikely to wander off at this age. Simply lead your mare out of the field and you will find that the foal will follow very closely. The reason I always advocate teaching the foal to lead is to avoid unnecessary accidents. It is so easy for a frisky young foal to slip over on wet concrete or find himself head-on with what you would consider a very normal yard obstacle. However, if you are on your own and you need to bring in your mare and foal, with any luck he will gladly trot alongside his mother.

Feeding the mare

As the mare spends more time out at grass, so feeding will change accordingly. The chart on p83 shows three feeds per day after birth for a horse mare, which can be reduced to two feeds per day when she is on good spring grass and is staying out for long periods.

Once the mare is out day and night, it is unlikely that you will have to give any extra hard feed, but this will depend on the condition of the mare and the grazing available. If in doubt about feeding, consult your vet or a knowledgeable stud person.

Feeding the foal

During the first six to eight weeks it should not be necessary to feed your foal; the mare should have sufficient milk to meet his requirements. The foal will soon become inquiring, and by the time he has reached two to three weeks old he will be picking at hay and grass, as well as nibbling from his mother's feed. If there is a valid reason for not permitting the foal to have any hard feed, for example in such cases where a vet has advised to keep the foal's weight down, then the mare should be fed from a manger over the stable door or somewhere that the foal cannot reach.

The only time that you will need to feed a foal at this age is when the mare is short of milk.

'Creep feeding' describes the method of feeding a foal where the feed is placed in a special compartment, too small for the mare to reach into, which therefore allows the foal to consume his own particular feed ration without interference. Today, feed manufacturers produce 'creep feeds' specifically formulated for foals from about the age of three weeks. The quantity required by each foal can only be determined by the prevailing circumstances.

FOAL ILLNESS AND ABNORMALITY

Keen observation during the foal's first three weeks will help you get him through this vulnerable period. Part of good foal handling is to know when to take action and how to deal with ailments that can arise. The ailments noted below should be read carefully before your mare foals, so that you are able to recognise any one of the symptoms described, and take prompt action if a problem occurs. It is **very important to get to know your foal's drinking habits** – the reason being that if there is something wrong, often he will be reluctant to suck (as you will note in the following four ailments). The mare's udder will quickly become full and hard and milk will squirt out as she walks around. The sooner you call the vet, the more chance the foal will have of survival.

Foal ailments

Diarrhoea in young foals is often called **scouring**. It is a very common ailment and can occur when your mare first comes in season after foaling – this first season is called the foaling heat and occurs between 7 – 12 days after birth. It is interesting to note that many breeders used to expect the foal to scour at the same time as the mare had her foal heat and it was believed that her milk changed its consistency; however, scientific evidence now shows that the mare's milk does not change during this period, and it is known that scouring can occur as a result of worms. It is therefore advisable to worm your foal at about seven days after birth, and every three to four weeks thereafter until he is three months old (see Recommended Worming Plan p111).

Immediate action must be taken if the foal is seen to scour. First take his temperature, and if it varies from the norm (see Temperature Chart p107) then call your vet. Meanwhile you can administer 20cc of any proprietary medicine for the relief of diarrhoea (such as, for example, kaolin and morphine available from your local drug store or chemist) by syringe into the mouth twice or three times daily. If scouring persists, call your vet.

Make sure you continue to observe: as droppings become normal, *immediately* reduce the medicine you are using. **Make your foal comfortable. Clean off the diarrhoea under his tail** with cotton wool and warm water **twice daily**; thereafter apply vaseline or baby oil – scouring burns, and this helps prevent loss of hair.

Infectious scours: Some scouring can be dangerous; if you notice pure liquid expelled quickly, take prompt action as foals dehydrate very rapidly – **call your vet immediately**. If there is more than one foal in the yard, wear overalls as it is likely that the condition is contagious. The foal will probably run a temperature, but even if his temperature is normal, I cannot stress more

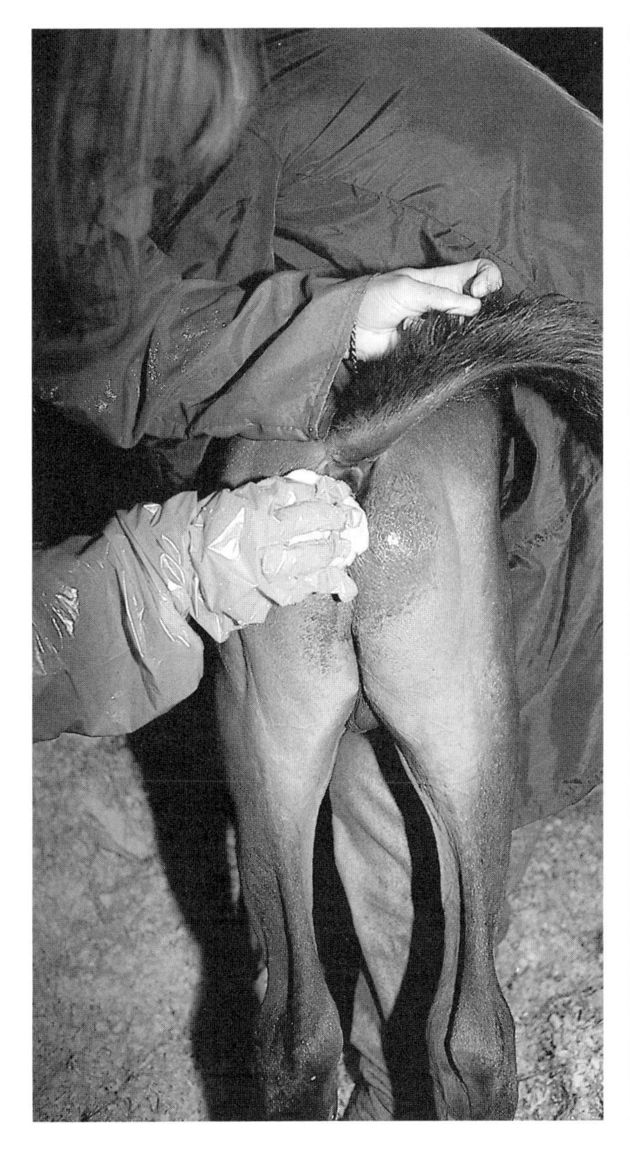

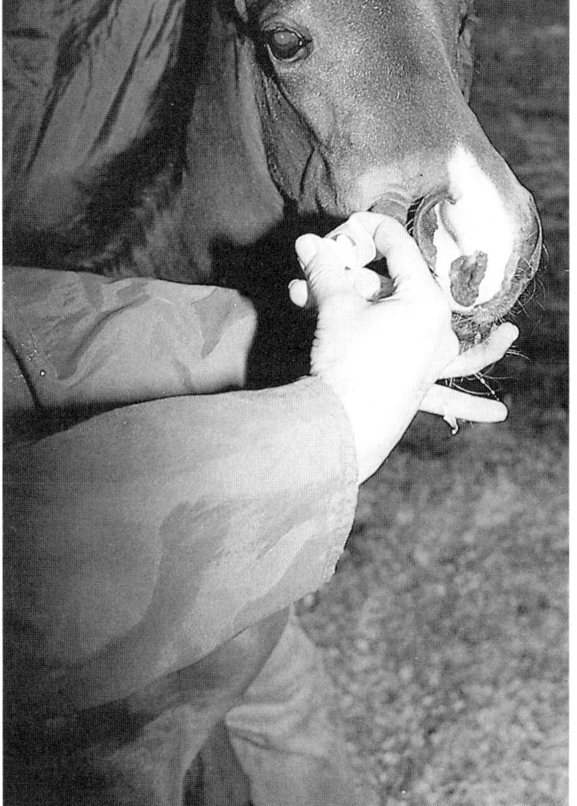

(Above) *Cleaning a scouring foal with cotton wool and warm water*

(Above right) *Worm the foal at seven days old and this will help to prevent scouring*

strongly, *call the vet*. **A dehydrated foal can die**.

To check for dehydration, take a handful of skin on the foal's neck and hold it tight for a few seconds; when released it should go back to normal quite quickly. If it does not, then your foal is probably dehydrated. A small amount of electrolytes mixed with 20cc of warm water can be syringed into the foal's mouth to assist in replacing lost salts, and repeat this every half hour, if he will take it; but if your vet can attend quickly, then it is best to await his instructions. Make sure that when your foal has recovered, no other mare and foal use that stable until it has been completely cleaned out and well disinfected.

Joint ill is an infection that enters via the navel – hence the reason for instant preventative treatment of the umbilical stump after birth – and usually occurs onwards from three to seven days after birth. A foal who for some reason does not receive sufficient colostrum is most at risk.

The foal becomes lethargic, disinterested in suckling, and probably lame – one or more of his joints may swell up. This swelling appears similar to a leg injury, but in fact is the result of an infection normally originating from the blood. The foal will also run a high temperature. If you notice any one of these signs, **call your vet immediately**: if treated *promptly* with antibiotics your foal may recover. (Recent research has resulted in new drugs which are now available on the market.)

Neonatal maladjustment syndrome (nms): Wobbler foals, barkers, dummies, sleepers: these are all names that you may have heard, and they relate to neonatal maladjustment syndrome. Foals who are affected by nms suffered a temporary shortage in the supply of oxygen to the brain during birth. I explained how important it is to leave the newly born foal quite still next to its mare before the umbilical cord breaks: if the cord is broken or breaks prematurely this could cause a shortage of oxygen resulting in nms. During the first two or three days of the foal's life there are various points to look out for:

1 If the foal stops suckling and loses its vital sucking reflex. **Call your vet**.
2 If the mare's udders fill up rapidly and perhaps milk runs off. **Call your vet**.
3 If the foal wanders round and round the stable appearing not to see the walls or his mother, as though he were blind. **Call your vet**.
4 In severe cases the foal may make a noise like a dog, hence the term 'barker'. Whatever the symptoms call your vet immediately; he may be able to save your foal.

It should be noted that the majority of severe cases are found in the Thoroughbred breed of horse.

Umbilical hernia: A hernia is a protrusion of bowel through a natural opening. If a swelling is noticed around the foal's naval where the umbilical cord was attached, it is highly likely to be an umbilical hernia; it can appear at any stage during the foal's first six weeks. Often the swelling will resolve itself as the foal matures, but if it is larger than ¾in (2cm) in diameter, I suggest you ask your vet for advice as to what action should be taken. In minor cases the sort of rubber ring (or rings) normally used for docking lambs' tails is

An umbilical hernia, before a rubber ring is applied

Nasal discharges are quite common during the summer months

carefully applied to the skin and close to the abdominal wall over the hernial sac whilst the foal remains standing. After three to four weeks the tissue below the ring will shrink and drop off. Ensure that your foal is vaccinated against tetanus before treatment.

In cases where the defect is too large for the application of rings, then an operation under general anaesthetic will be required. Your vet will discuss whether this operation should take place before or after weaning.

Runny noses or nasal discharges: These are quite common in foals. Unless symptoms occur in the first three weeks of a foal's life (which is unusual) they are best left to run their course. *However*, if you notice **any signs of listlessness** or if the foal **runs a temperature**, *take action* and *call the vet.* **Plenty of fresh air is advised**. You should clean off the nostrils with cotton wool and apply a decongestant cream to help clear the nasal passages. Snotty noses can continue for some weeks and even if the vet prescribes antibiotics, the condition often re-occurs after the course is finished. There is little you can do to help, but **observe** and keep your foal either out at grass or if in at night use a stable where there is plenty of fresh air circulating.

There are many other foal ailments, but they are not so common and therefore not listed here. Treatments for the above ailments should be obtained from your vet and more detailed descriptions can be found in veterinary books. If you are worried about any unusual circumstances *do not hesitate*, consult your vet.

USEFUL INFORMATION RELEVANT TO NORMAL FOALS

Temperature: 37.3C–38.3C (99F–101F).
Pulse: 100 beats per minute. At birth 80 beats per minute; rising to 140.
Respiration: 30 to 40 per minute.

Foal abnormalities/deformities

Cleft palate: A cleft palate is quite rare, but it is very easy to recognise. If you notice signs of milk running from a foal's nose when he drinks, it means that there is a small gap or split in the bones of the roof of the mouth which allows a proportion of the milk to escape down through the nose. This condition is known as a cleft palate and in mild cases surgery can rectify the problem.

Limb problems: There are several variations of limb problems in newborn foals. At first sight your newborn looks wonderful, and so pretty as it follows its mother out to the field, and you observe those delightful first few minutes as it scampers around. However, it is after this that you might see any limb problems more clearly. Many foals are born with legs that are not completely straight, more often noticeable in the front legs. The majority will right themselves, given time, but if the deformity is severe or if you see no improvement in the first few days of the foal's life, then seek advice from your vet.

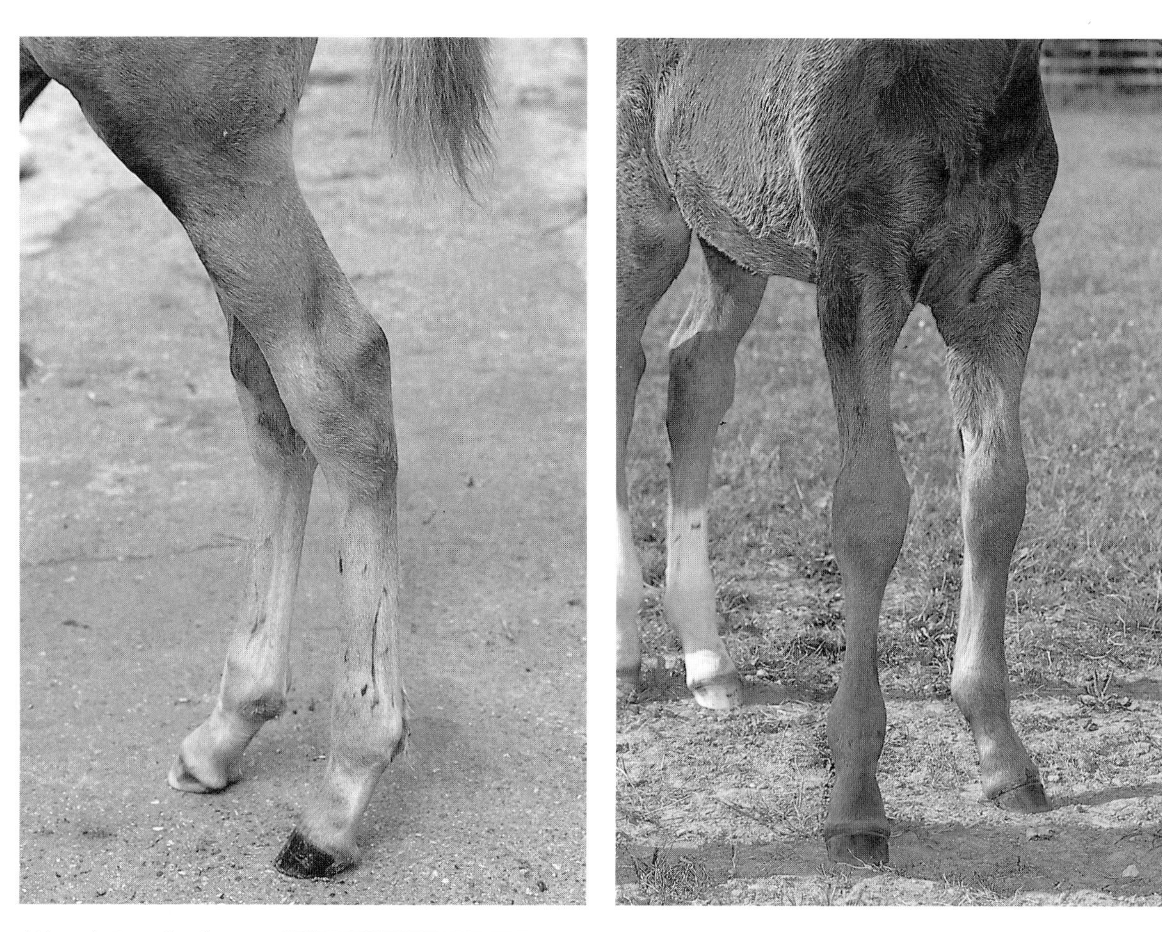

(Above) *Poor hind legs, a fault of conformation; the limbs will strengthen, but the poor shape will remain*

(Above) *An example of poor forelegs. Note how much the foal is over at the knee on the near side*

(Right) *An example of correct forelegs in a foal*

Angular limb deformities: There are a number of different deformities, the most common being known as *carpus valgus* (splayed front legs). The causes are numerous, but the main reason is that growth occurs more rapidly on one side of the leg than the other, making the leg unbalanced and subsequently crooked. Uneven weight distribution across the joint can also be the cause of limb deformities. The reasons are scientifically complex and not in the boundaries of this book, but it should be stressed that the importance of correct feeding prior to birth, together with the appropriate mineral supplement and correct calcium/phosphate balance are important.

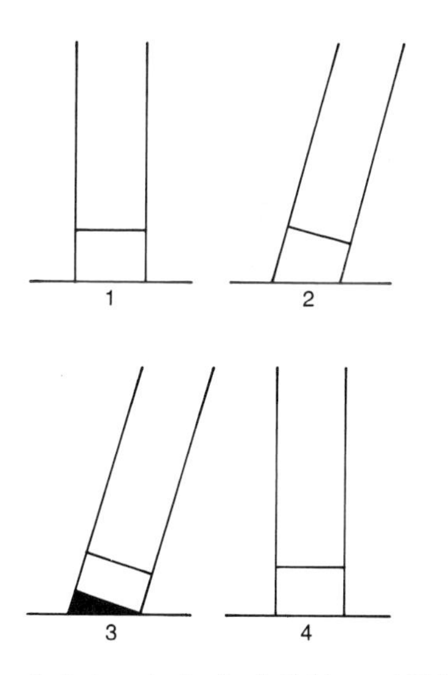

Correcting leg deviations in the foal. 1) Normal 2) Deviation of the lower leg due to uneven bone growth 3) Shaded area indicates where the hoof needs to be trimmed to make the foal bring the leg back to the vertical 4) The corrected foot

Contracted tendons: Contracted tendons are quite common in new-born foals, especially in the forelegs when the foal appears to be walking on his toes. In fact the tendon is not actually *contracted*, but due to the fact that the foal lies in the mare's womb with his legs in a flexed position, when he is born the tendons a) may not have kept up with the rate of growth of the canon bone, and b) it takes a while for them to develop and become 'springy'. The problem normally cures itself as the foal takes exercise and gains in strength, but if the condition is severe, massage or strapping up, most likely with a splint in place for a couple of weeks, may be recommended by your vet. However, if you do use a splint, remember that the skin of a foal is very soft and vulnerable so the dressing must be changed daily and good padding used.

Hyperextension of limb bones: This is the name given to a foal born 'down on its fetlocks' (see Case Study 1 p84). In severe cases, as the picture shows on p84, the pastern should be lightly bandaged, placing gamgee over the heels for protection. Make sure that the bandage is not so tight as to restrict movement. With limbs such as this when the foal is about three weeks old it is likely that the vet will advise the use of plastic glue-on shoes with a small heel extension. (I have attempted to use ones with a longer 6in (15cm) heel extension, but had little success in keeping them on, and they give such a steep lift that it tends to be too great a strain on the pastern.) Plastic glue-on shoes are not easy to apply and are expensive. I advise using a blacksmith who is already familiar with the fitting of foal shoes – though be warned, it will take him about one-and-a-half hours to fit a pair of hind shoes. In the case of Josie's foal (p85) the shoes were left on for three weeks, and then changed on two further occasions only. It is of paramount importance that you do *not* leave shoes on a foal for longer than a three-week period as they will start to restrict normal foot growth.

Parrot mouth: This is an **overshot jaw** – in German the literal translation is 'over-bite'. The teeth of the top jaw are seen to protrude over the lower teeth, and it is an hereditary conformation fault. There is nothing that can be done to correct the deformity, but it rarely causes any problems. The protruding amount could be as little as 2mm (1/10in) or as much as 4cm (2in); in severe cases the foal may have difficulty in masticating hard food, and as a result could drop in condition once he is weaned from his mare. A close eye should always be kept on the well-being of such a foal. The less severe cases sometimes change for the better during their first two years, resulting in a normal mouth at 3 years old.

An **undershot jaw** (also known as sow's mouth) is very rare.

Physitis (formerly referred to as Epiphysitis): This is the name given to abnormal bone growth (activity in a growth plate). It is most common in yearlings *but* can occur in a large foal with a rapid growth rate (see p136).

Sickle hocks: These can look very prominent at birth, but with time the hocks will strengthen so that as the foal grows the sickle shape will lessen. However sickle hocks are a conformation fault, and are unlikely to disappear completely.

A severe case of a parrot mouth, which nevertheless caused the foal no problems in masticating his feed

Splayed front legs: Don't worry, the front legs will almost certainly become completely straight, *although* every foal is different. Do not hesitate to consult your vet – he may well suggest a course of action that will help; but I would wait until the foal is about two/three weeks old before calling a vet (see p109, Angular limb deformities). In some cases the hoof will require trimming at frequent regular intervals.

Upright pasterns: Some foals are born with very upright pasterns; once again, only time will assist in rectifying this, but I suggest you keep the foal out as much as possible and make sure that he

2 wks before birth — 2 wk after.

does not have access to any extra feed. He should not be allowed to become overweight, although obviously the mare's milk should never be restricted, unless recommended by a vet. If a foal is still over at the knee at about three weeks old consult your vet.

Teeth

In the course of your 'stringent' observations you will no doubt be surprised how soon your foal's first teeth show through. It is interesting to note that either at birth or more likely during the foal's first week, the two milk teeth known as **central milk incisors** will appear. Between 4 to 6 weeks after birth the **two lateral milk incisors** will follow. The teeth then remain the same until well after weaning (approximately 9 months) (see p147).

WORMING PROGRAMME

Worming control

Routine worming is of paramount importance to the well-being of your foal, and I cannot stress too strongly *how vital* it is to carry out the recommended procedure. Your foal may not appear to be suffering at the time, and you may be of the opinion that your paddocks are free from worms, *but* believe me you can *never* be sure. It is only later on in the life of a youngster that the resulting damage will be discovered.

The mare should have been regularly wormed (see p47) during pregnancy and shortly before giving birth (two weeks before is recommended). It is normally suggested that you give your first wormer to the newborn at four weeks old, *but* recent research shows that it is possible for the foal to pick up **threadworms** (*strongyloides westeri*) from the mare via her milk.

A heavy infection of threadworms can cause scouring (diarrhoea), which may well coincide with the mare's foaling heat (first season) approximately ten days after birth. For this reason I suggest a preventative worming for your foal with ¼/½ a syringe of, for example, Panacur paste (UK) (according to the size/type of foal) **at seven days old** (see Foal Worming Plan).

TYPES OF WORM THAT AFFECT THE FOAL

Threadworm (*strongyloides westeri*)
Large roundworm or white-worm (*ascarid*)
Large strongyles or large redworm (*strongylus vulgaris*)

Ascarids are often prevalent in large numbers in young foals. This particular species does not cause problems in yearlings or older horses as the youngster builds up a resistance against them. *Ascarids* are the worms that often cause a foal to have a dull coat and pot-belly. They are between 2–4in (5–10cm) in length, and foals can have up to 1,000 adults in their gut at any one time. What happens is that the eggs are passed out by other foals/horses, and as they have a sticky outer layer they adhere well to all objects, including grass; the foal then eats them together with the grass. When the eggs find their way to the intestine, the larvae hatch and pass through to the liver and then to the lungs, where they are coughed up and swallowed again. Basically throughout this process they are absorbing the foal's prescribed nutrients and hampering his development.

Apart from good pasture management – which, unless you are in the lucky position of owning land, you can do little to assist – the only other ways in which you can help are to make sure that the stables where the mare and foal are kept are regularly disinfected, and that your mare is wormed thoroughly after foaling (see p112). Knowing that foals are susceptible to these worms, may I suggest that you follow a simple worming programme such as the one below, or ask your vet for his opinion on an effective campaign against worms.

Foal worming plan

			Pony foal	Horse foal
			*Approximate measure	
1st dose	7–10 days	Panacur paste	¼ paste	½ paste
2nd dose	3 weeks	Strongid-P paste	¼ paste	½ paste
3rd dose	6 weeks	Panacur paste	¼ paste	½ paste
4th dose	9 weeks	Eqvalan paste	¼ paste	½ paste
5th dose	12 weeks	Strongid-P paste	¼/½ paste	1 paste

Then worm *every four weeks* **until the foal is at least 8 months old, or preferably** *one year*.

Consult your vet if in doubt as to what quantity of wormer to administer. Once the foal is more than ten weeks old, some worming pastes are often administered in larger quantities, *but* you should consult your vet beforehand.

After 12 weeks you should increase the time in-between wormers to every four weeks until the foal is 8 to 12 months old; thereafter see Youngster Worming Plan p149.

Mare worming plan

After foaling, worm the mare within the first two weeks, and every four to six weeks thereafter.

VACCINATION PROGRAMME

Your mare should have had her annual vaccination boosters in the latter half of her pregnancy (see p48). In the UK a combined vaccination known as **Prevac-T** (equine influenza vaccine combined with anti-tetanus) is used. In the USA one should consult a veterinarian for a vaccination schedule; remember that your mare should have received rhinopneumonitis vaccines during her pregnancy. The proper vaccination programme will ensure that the mare's colostrum has a maximum level of antibodies to act against both equine influenza and tetanus to protect the foal. If you live away out in the country and do not vaccinate as routine, then I **strongly recommend that she has an anti-tetanus vaccine at least one month before the expected birth**.

It is important to know when the mare was vaccinated in order to advise the relevant programme for your foal. *Providing* the proper vaccinations were given to the mare, as discussed above, your foal need not be vaccinated until he is three or four months old. However, there are two exceptions to the norm:

1 if you have an orphan;

2 if your foal is wounded, the vet may decide to give an anti-tetanus as additional immunisation.

I suggest that if for some reason your foal is not vaccinated at three or four months, then he should be done *prior to weaning* so that he has ample protection once he is no longer suckling. It is *not* advisable to vaccinate your foal at a time when he is perhaps unwell with, for example, a summer cold.

As an example of a vaccination schedule, in the UK Prevac-T is given in three parts:

	Not less than	Not more than
1st dose	3 months old	recommended 6 months old
2nd dose	21 days (3 weeks) later	45 days (6 weeks) later
3rd dose	150 days (5 months) later	215 days (6–7 months) later

This vaccination must then be repeated within 12 months, ie **annually.**

Warning: If you do not vaccinate *within the year* you will have to start the course all over again, which is an expensive exercise.

When your vet comes out to vaccinate, make sure that he completes the identity of the foal on your certificate; this should be passed on to anyone who purchases him at a later date.

Some shows today ask to look at your vaccination certificate before allowing you to enter the showground. Occasionally your foal may feel a little 'under the weather' after the vaccination so I do not advise having it done, for example, the day before a show.

REGISTRATION

Most breed societies require notification of birth within 28 days. Now that you are the proud owner of a beautiful healthy foal, remember to check what is required for registration and send off the relevant paperwork.

HANDLING AND THE FARRIER

Handling and observation should continue throughout your youngster's foalhood and his adolescence. **But do not overdo it.** Use your common sense, as even the most long-suffering foal can get fed up with fussing. Titbits are definitely *not* recommended. A foal likes to nibble at this

Practise picking up the foal's feet in preparation for the farrier

age – do not encourage it. Treat him in a quiet way, like an adult horse – for example if, when you walk into the stable, he shoves up to the door, just gently push him back; manners count even at this age. Until such time as mare and foal are out at grass day and night, you will be handling/leading the foal on a daily basis – and that is enough at this time, except *do not* forget his feet.

The farrier will be needed in the future, so teach your foal to pick up his feet when asked: have someone to hold his head and talk to him while you run your hand down each leg so that he gets used to the feeling (foals can be quite ticklish on their legs). Then gently hold up one foot at a time for a few seconds – and put each one down when *you* want to, not when the foal does; with patience, he will learn to pick up each foot on command. Don't forget to use your voice as you would with an older horse.

If, for whatever reason, a foal has not been taught to pick up his feet by the time he is about eight weeks, then you may find the exercise quite difficult and some foals will behave in a stroppy manner, especially when asked to pick up a hind

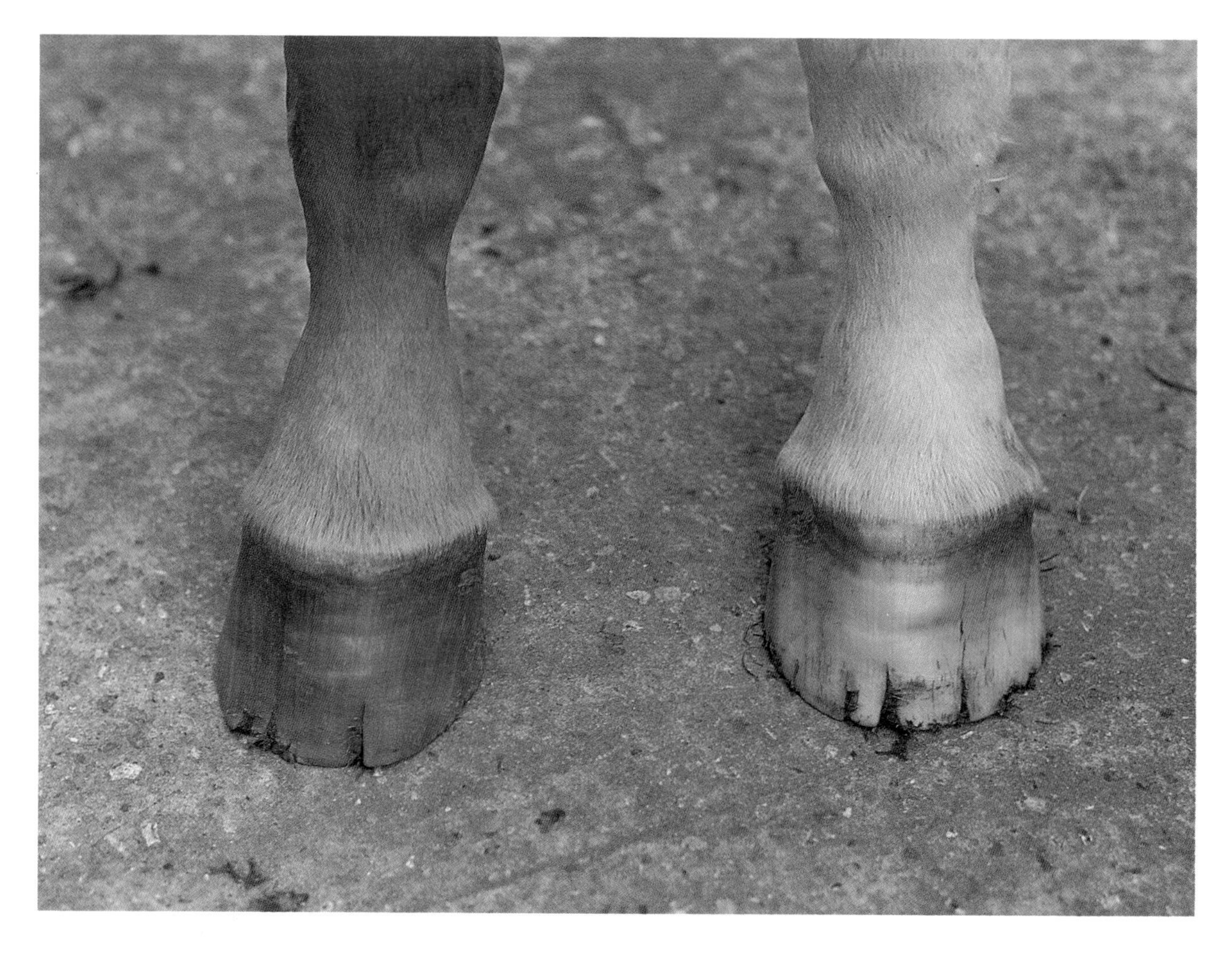

A foal's foot that has been neglected

foot. It is easier to practise in a stable where you have a wall alongside which you can stand the foal; and you will also need someone quite strong to assist. The secret with the hind legs is to pick up one leg and keep hold of it even if he is kicking it back and forwards; don't lift it up too high, and eventually he will hold it still – and you will have won your first battle of wills!

The farrier's visits

So often feet are neglected – maybe it is just ignorance, possibly it is a short-sighted way of saving the farrier's bills. But remember the old adage 'no foot, no horse': nothing could be more true, so be diligent, and arrange to have your foal's feet trimmed on a regular basis. I suggest the first trim should be carried out when he is 4 to 6 weeks old.

Exception to the rule: if your foal has a faulty foot position or some other leg deformity, however slight, ask your vet or your farrier to have a look. Walk the foal up in a straight line on hard ground for the inspection. Often the farrier can greatly assist in correcting the problem if he is called in at an early stage, and at frequent intervals.

Be consistent with your organisation of the farrier's visits – regular *monthly* trimming up until the end of his first year is recommended. The foal's foot may not appear long, but remember this is the formation growth period and it is of paramount importance to keep the foot in a good shape at this time.

SUMMARY:
HOOFCARE FOR YOUR FOAL

First trim: 4–6 weeks old (unless vet recommends otherwise).
Second trim: 10 weeks: now continue on a monthly basis.

Hoof oil/Hoof grease

Is this a necessary addition to your list of tasks? It shouldn't be: provided the foal is getting the correct feed from his mother, the hoof should naturally grow down good and strong without the need for any external applications. However, if you are taking him showing, then hoof oil should be applied for cosmetic purposes.

If the foot is neglected and not regularly trimmed, then splitting will almost certainly occur, because the foot will break up on the ground. In this instance use a hoof treatment product on a daily basis which should speed up hoof growth and assist the hoof in recovering its original elasticity and form.

IMPORTANT POINTS TO REMEMBER FOR THE BENEFIT OF YOUR FOAL'S FUTURE
Regular worming Up-to-date vaccination programme Complete registration papers, and send them off Regular visits by the farrier

IN-HAND SHOWING OF MARE AND FOAL

In-hand showing takes up a great deal of time, but when done well, it certainly benefits the foal as regards discipline and early learning of manners. Furthermore there is no doubt that if you are a 'professional breeder' or a single mare owner wishing to sell your youngstock, in-hand showing is the best way to promote them. *After all, you do not have to be a good rider to show off your stock in hand.*

Showing can also be a very rewarding exercise, though before you embark on an outing with your mare and foal, do make sure that you know exactly what is expected of you in the show ring. You and your horses must be turned out correctly: irrespective of how small the show, it is important to show off your stock well, and their turnout reflects on you. The basic needs will be noted here but fundamental requirements in addition to these are **common sense** and **forethought**.

A lesson in leading the young foal

Travelling arrangements

The lorry or horsebox: All horse-knowledgeable people know that when travelling horses a 'safe' lorry is of the utmost importance; with a mature, calm horse which travels well, you can often get away with a flimsy partition or a sub-standard interior, but it is **not worth the risk when you travel a mare with a foal at foot**. Your mare should travel tied up and her foal loose by her side, and like this, if your lorry is not safe, he can get into all kinds of trouble. **Make sure** that:

1 the interior is safe, with no protruding objects;
2 there is no gap at the front of the partition where the foal could climb underneath;
3 the ramp is not too steep, or the step off the ground and into the lorry is not too high.

Bed down the area where the mare and foal will travel with a thick layer of straw and bank it up along the edges, so that if the foal lies down he will be quite comfortable.

Travelling in a trailer: There is no reason why mare and foal should not be comfortable in such a vehicle, provided that the middle partition is removed, and that normally the mare travels quite happily in a trailer. I suggest tying her up so that she cannot turn round – it is probably safest to rack her (put on two ropes, one each side of her headcollar, tied quite high up); this will allow the foal to walk around the mare without any trouble.

Make sure that the side door (if there is one) is locked or tied up securely, so that if the foal leans against the handle it will not fly open. The trailer should be well bedded down with straw. One advantage of using a trailer instead of a lorry is that the ramp is not as steep and is therefore much easier for loading and unloading a foal.

Equipment required

For mare and foal: This varies depending on the breed of horse or pony. In all cases the mare should wear a bridle. Specialist in-hand bridles are always correct, but for a hunter mare a double bridle is the accepted uniform. With an in-hand bridle a leather lead rein is used, with a double bridle the two normal leather reins. For smaller shows it is quite acceptable to exhibit in an ordinary snaffle bridle, providing it fits correctly and is fitted with a **cavesson noseband** (as opposed to a drop or flash).

The foal needs a well-fitting slip or headcollar, depending on his size. This should be made of leather; at a small show you could use a dark nylon headcollar, although it is not 'correct' to do so. The leading rein can be either leather or white tubular web with a leather fitting and buckle. **Note: Foals under the age of six weeks should not be shown** (most schedules stipulate this rule).

For handlers: There are no hard-and-fast rules and regulations relating to the handler's attire, although more recently some breed shows (notably the Warmblood and Trakehner shows) do specify certain wear. Traditionally a jacket and tie with straight slacks or jodhpurs and jodhpur boots for gentleman and lady are correct. At county shows the lady should wear a hat, and the gentleman a bowler. Handlers are often seen to wear jodhpurs with long boots, but these are useless if you find it difficult to run in them! Gloves should always be worn and a cane carried. It looks smart for the two handlers of mare and foal to be dressed alike.

At Warmblood shows it is currently the trend to wear white or black slacks with a white shirt, as is the norm in Europe.

Turnout of mare and foal

The mare: It is important that your mare looks well and in good condition physically. It is perfectly feasible to take a mare out of the field, give her a bath the day before, stable her that night, and plait her the next morning – if she looks great, you are in luck, and depending on the time of year, this may be possible. However, if she looks under-weight, has evidence of fly bites or flies' eggs and her feet have not been trimmed, do not take her to a show. To show most mare you need preparation time: she would be stabled at night for at least two weeks before, maybe longer, depending on her type and condition.

To present her correctly in the ring, she should

have her mane plaited (unless she is in the mountain and moorland class). The plaiting of tails is not compulsory. It is a good idea to practise plaiting her up some time before the show to judge if she looks best with small or large plaits; this will depend on the shape of her neck and her conformation. Obviously she should be well strapped and groomed, with feet oiled and baby oil or similar applied round the eyes and nostrils. White socks should be washed the day before and whitened with a chalk block or powder before bandaging.

The foal: You do not have to plait your foal's mane. I prefer to see foals with unplaited manes, but if they are over five months, the mane – and likewise the tail – can look messy and may need to be plaited to look neat and tidy. It is not advisable to pull a foal's mane at this age as it is too painful for them. Make sure that your foal is used to having his feet oiled so that if you want to do this on the showground, he will stand still. Like his mother, he should be in good physical condition, well groomed and turned out.

Preparation: training and appearance

Considerable preparation is needed if you decide to show mare and foal. You must be prepared to spend some time on training, and you will need an assistant – someone to lead the mare while you train the foal. In the ring you will be expected to lead the mare, with the foal following behind, around the arena with all the other competitors. All exhibits are then brought into the middle of the arena, and each one is asked to stand forward individually for the judge to inspect.

It is normal to judge mare and foal separately; whichever is being judged, you will have to keep the other one out of the way. Generally, you are then asked to walk away from the judge in a straight line and trot back. If the foal is being judged it is quite in order for you to trot your mare up in front of the foal so that he follows willingly. As you do this, you will have to keep an eye on what is happening behind you; for example, if the foal breaks into a canter, slow the mare down until he starts trotting again. *Use your common sense* – be alert, and handle the mare so that everything is made as easy as possible for the foal and his handler. There is an art to trotting up any horse that is being shown in-hand – it is not easy and requires a great deal of practice (see p151).

I advise a lot of practice at home; be patient, it will get better. Be firm with your foal, always use voice commands, and carry a stick to assist – this is used merely as an aid to back up your voice, not to beat him up! Make sure that you *always* walk alongside him, *never* in front. At first you will probably find that your foal will not walk away from his mother; however, follow the suggested lessons (see box, p117), and gradually he will have

EXERCISES TO PRACTISE BEFORE GOING TO YOUR FIRST SHOW

1. Walk mare and foal together in straight lines, sometimes with the foal at her side (offside and nearside), and at other times behind. Make sure they walk out well and do not slop along or dawdle.
2. Repeat this exercise at trot.
3. Stand the foal still and walk the mare away from the foal a few yards; then allow her to join him. **Use voice aids at all times** and **do not forget to praise him when he does it right**.
4. Repeat this exercise, standing the mare still and taking the foal away; first at walk and then at trot.
5. Practise walking the foal in front of his mother – he may not be keen to do this, so use a positive 'walk on' voice aid and back it up if necessary with your stick. As the mare is right behind, her leader should be able to help by pushing him on as a reminder from behind.
6. You need to practise every other day for two to three weeks. Use different areas, and beware of always practising on hard ground – sometimes a foal will get sore feet, so find a field with good grass, or an artificial surface.
7. Once you find your trainees are doing well, practise all these exercises in the company of another horse or two if you can – though make sure it does not come too close to your foal-proud mare.
8. Occasionally practise leading your mare in her bridle, as she just might react in a different way when bridled up.
9. Ask someone to act as a 'pretend judge': stand the foal up so this person can walk around him, run a hand down his front and hind legs, and generally observe him as a judge would. You will be surprised how shy foals can be, and yours may well not like this at first.

Loading a young foal: two people join hands behind the foal's hindquarters. The mare may be less upset if the foal is led up the ramp in front of her

the confidence not only to walk away from her, but to show off his paces in the correct way, too.

Finally, **do not overdo your training** as both mares and foals tend to get very fed up if you mess around with them too much. It is up to you to differentiate between plain naughtiness and the beginnings of a sour foal.

Loading practice: It is debatable whether it is a good idea to practise loading and unloading. Assuming that your mare loads well, my answer might be no; especially if travelling in a lorry when you could well give the foal an unnecessary fright. However, if you are using a trailer, it can certainly do no harm to practise, providing you have three people: one to lead the mare, one to lead the foal, and one to help push the foal in behind the mare and then to quickly put up the ramp.

Loading a foal can be quite difficult, even if the mare loads willingly. First, take the mare up the ramp and ask the handler to hold her in the box, but do not tie her up at this stage as she might panic at being separated from her foal. Two people will probably be needed to get the foal up the ramp. If he is quite small, the two people can join hands behind his hindquarters and gently ease him up the ramp; it is always a worry travelling a mare and foal, but providing you take care and think ahead all should be well. Remember if the foal is very young you may have to bring him up in front of the mare. Otherwise she will get upset.

Unloading is also not so easy. The foal will almost certainly not have the courage to come off the lorry in front of the mare, but once she is unloaded, he may panic and be desparate to come off and join her – so inevitably he takes an enormous leap, which could well end in his falling flat on his face. Take with you someone whom you know is a **good handler**, who will restrain him

(Opposite) *A well fitting nylon foal headcollar*

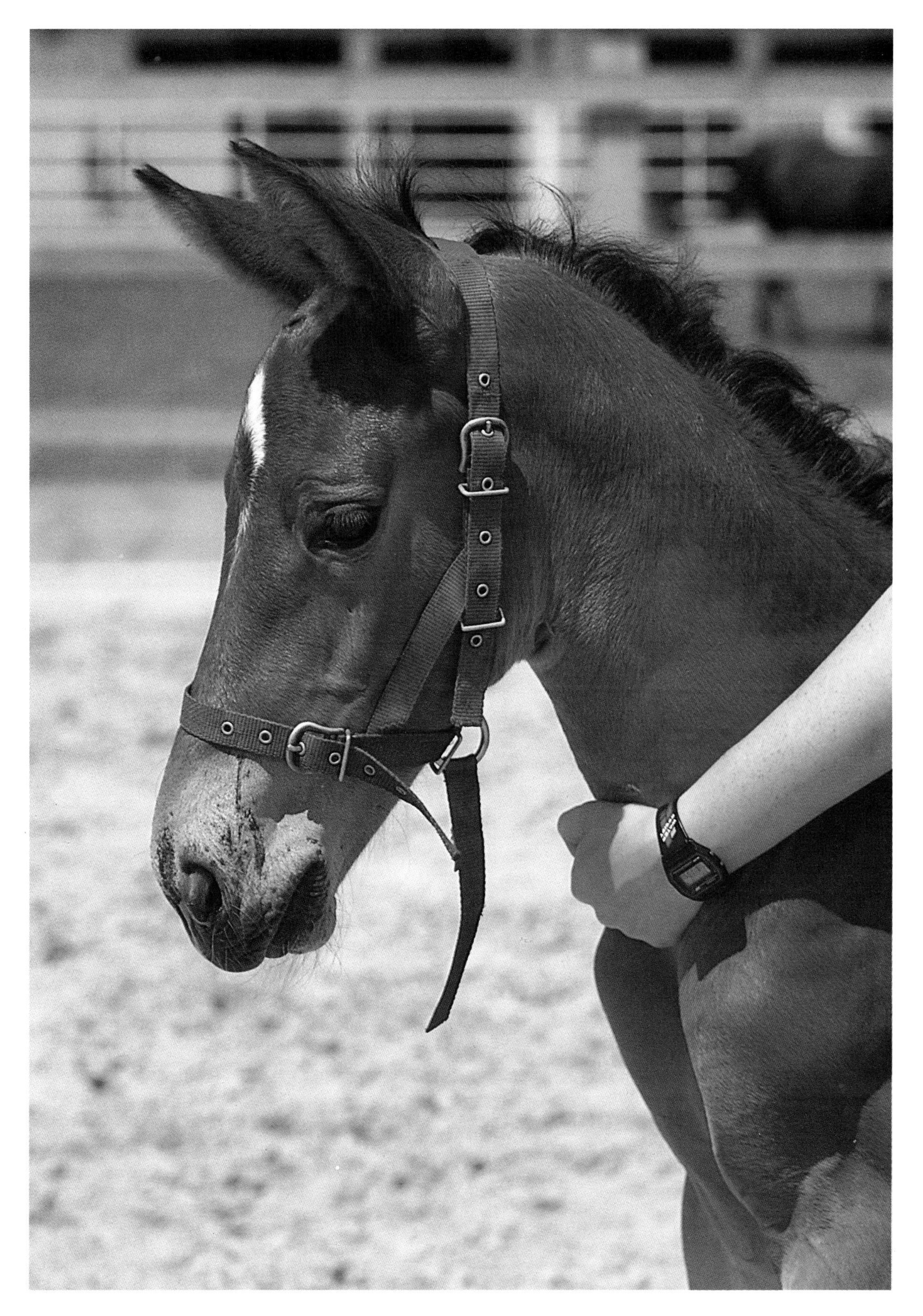

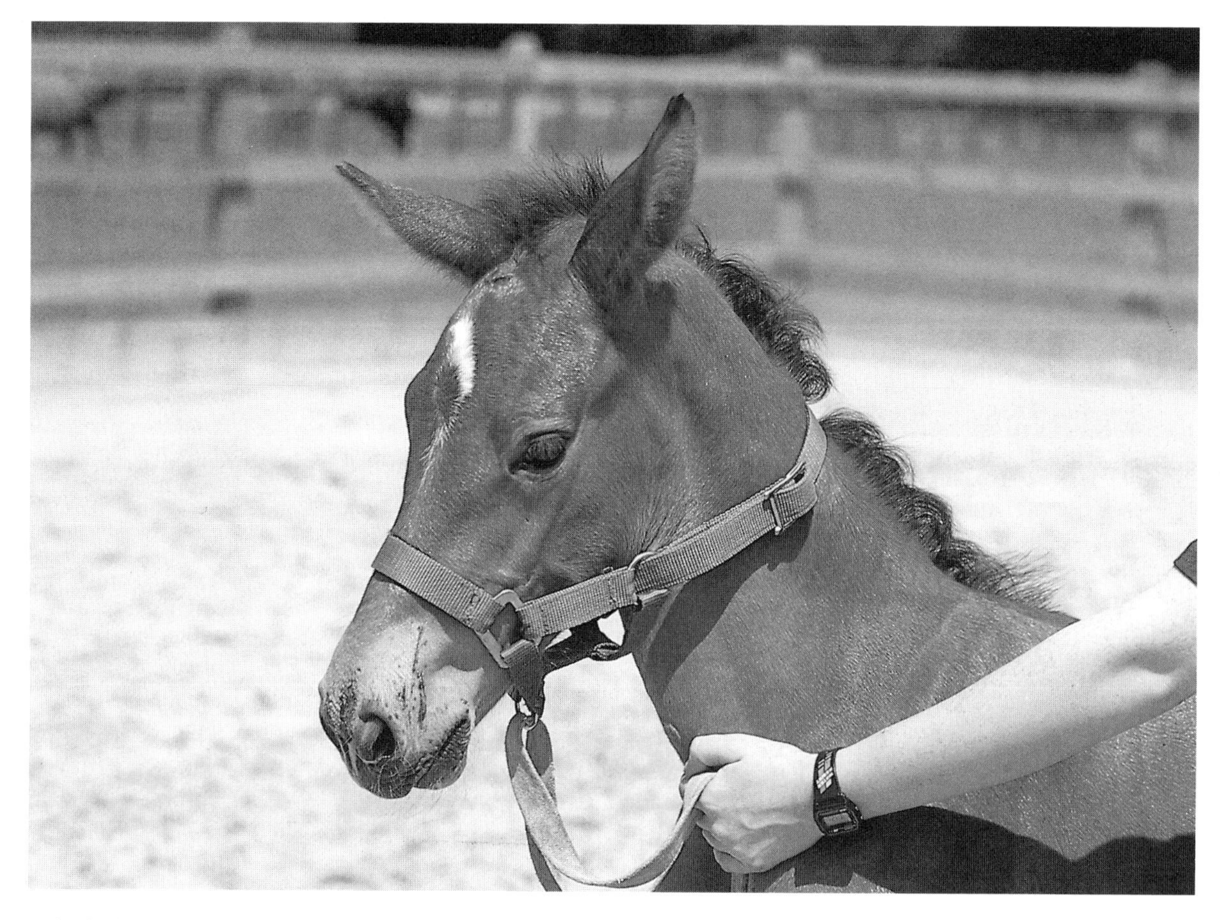

A badly fitting nylon foal headcollar

from leaping but will guide him off in a sensible manner. Always unload on the grass if possible, so that if the foal does leap, he is less likely to hurt himself. Also keep the mare very close to the ramp so that the foal realises that his mother is not going to disappear out of his sight.

Some foals have the opposite reaction and do not want to come down the ramp at all; they dig their heels in and will not budge. This situation also needs careful handling because if you push too much from behind, the foal could fall on his knees; so common sense and patience are required.

Bandaging a foal is not advisable. Their tiny limbs are not suited to bulky items and bandages could cause harm if done up too tightly. However, you may wish to protect the legs of the larger, older foal so in these circumstances small travel boots or light bandages can be put on. **Rugging up a foal is also unnecessary** unless this is specifically recommended by a vet in the case of a sick or an orphan foal.

Final appearance

The day before the show you decide to bath your mare – **do you also bath the foal?** If the sun is shining and it is a hot summer's day it will do no harm, but have an extra person to help, and remember this will not be a very pleasant new experience for your foal. Use warm water, and rub him down well after bathing so that he does not catch a chill. It is often easier to put the mare in the stable and stand the foal directly outside, under her nose.

And off you go!

On the show day itself leave plenty of time. You are well prepared, but make sure that you arrive at least one hour before your class begins so that the mare and foal can settle down in their new environment before going into the ring. Practise trotting up on the showground. Look alert and smart, and all should go well!

SHOW CHECK-LIST

Equipment for mare and foal (one day only)

Mare's show bridle	2 water buckets (without handles)
Lead-rein for above	2 feed buckets (without handles)
Foal's show headcollar	Feed
Lead-rein for above	Hay
Lunge-rein	Spare headcollar
2 grooming kits including:	Spare rope
Body brush	Plaiting kit including:
Curry comb	Scissors
Water brush	Tape
Stable rubber	Cotton
Sponge	Needles
Towel	Hairspray
Vaseline or baby oil	Summer sheet/cooler
Hoof oil and brush	Saddle soap and sponge
Chalk block or powder	Vaccination certificate
Brass cleaner	
Filled water container	

Equipment for handlers

2 show canes (or whips)	Trousers
Gloves	Boots/shoes
Jacket	Hat (not compulsory)
Tie or stock	Shoe polish and brush
Shirt	Clothes brush

Equipment for two-day show

Add:

Extra feed	Fork
Extra hay	Shovel
Feed manger	Night rugs
Bedding	Shampoo
Broom	Show Sheen

HORSE HEIGHT CONVERSION CHART
(approximate equivalent)

hh		inches		cm
10 hh	=	40 inches	=	102 cm
12.1	=	49	=	102
12.2	=	50	=	127
12.3	=	51	=	130
13	=	52	=	132
13.1	=	53	=	135
13.2	=	54	=	137
13.3	=	55	=	140
14	=	56	=	142
14.1	=	57	=	145
14.2	=	58	=	147
14.3	=	59	=	150
15	=	60	=	152
15.1	=	61	=	155
15.2	=	62	=	157
15.3	=	63	=	160
16	=	64	=	163
16.1	=	65	=	165
16.2	=	66	=	168
16.3	=	67	=	170
17	=	68	=	173
17.1	=	69	=	175
17.2	=	70	=	178

7

THE WEANED FOAL

Is it time to separate your mare and foal? What does this really involve? Is weaning a major problem? Are your facilities sufficient? Should a colt foal be gelded at this stage? What about the mare – will she object to being separated from her offspring? She might well be carrying another foal, or you may wish to bring her back in to work. And the foal, does he need a playmate?

The word 'to wean' literally means 'to coax the youngster away from its mother and accustom him to food other than his mother's milk'. There are many different ways of doing this and some strong thinking on the subject: what I endeavour to put over in this next section is how important an influence the 'weaning period' may have on your youngster's future life. You could liken it to your child's first playschool – you have done well and got him through his vulnerable first few months, so keep up the good work! **The next period is an influential one** where quality of lifestyle and discipline will have a bearing on the youngster's future attitude to the world. Your financial position and available facilities will no doubt be restricting, but nevertheless, consider all options.

This chapter also looks at the questions of travel and purchase: for example, you may decide to take either mare *or* foal away from their present location, so how will you travel them? You may be considering the purchase of a young foal – where do you start looking? What should you expect when viewing: is a veterinary examination advisable, and when should a new purchase be brought home? These and other aspects of purchase, including the importance of studying bloodlines for breeding and performance, should all be considered before you part with your money.

WEANING THE FOAL

Is your foal ready to wean?

It is now five months since your foal's birth date and he is well grown and healthy. He has been regularly wormed, and has had his first vaccination. I am sure you will have observed how independent he appears to be, no doubt grazing some way away from his mother without any worries. I would even guess that those of you who bring your mares in at night may well have experienced the problem of persuading your foal to follow along behind – he is no doubt just as content to stay with his friends. However, if you are keenly observant, you will have noticed that if the foal perceives 'danger signs' – for example strange loud noises or a new horse in an adjacent field – he will quickly run back to his mother. This gives him renewed confidence to return to grazing. Maiden mares (ie mares with their first foal at foot) tend to be more protective, and at the slightest 'danger sign' the mare will go to the foal's aid.

A six-month-old foal is genuinely ready to wean, but if circumstances dictate, a slightly earlier separation, provided he has reached **five months of age**, can take place. If you need to wean a foal at *less* than five months, great care and attention to correct feeding and companionship will be required. **Avoid unnecessary stress** – the whole process of weaning is stressful anyway, so consider alternative methods carefully.

It is **also important to consider the circumstances of the mare**, too – is she in foal again, or do you want to put her back under saddle? If she is in foal, then for the benefit of the mare and to ensure that once again you give her what is best for her during the gestation period, when her foal is six months old it is time to wean him off. After all, she may well be four to five months pregnant and the unborn foal is growing fast. On the other hand, if she is not carrying and you wish to ride her then there is no hurry. Certainly it is more difficult to bring a mare back into work with a foal at foot, but it is feasible (see Gradual Weaning p128). If the mare is not in foal, no harm will be done if you leave the foal on his mother for another few months, but **be warned** – if you do not handle him he will not be easy to discipline later on. If the foal remains with his mother for a much longer period, he will eventually wean himself; however, a two-year-old still with his mother may well be observed suckling occasionally.

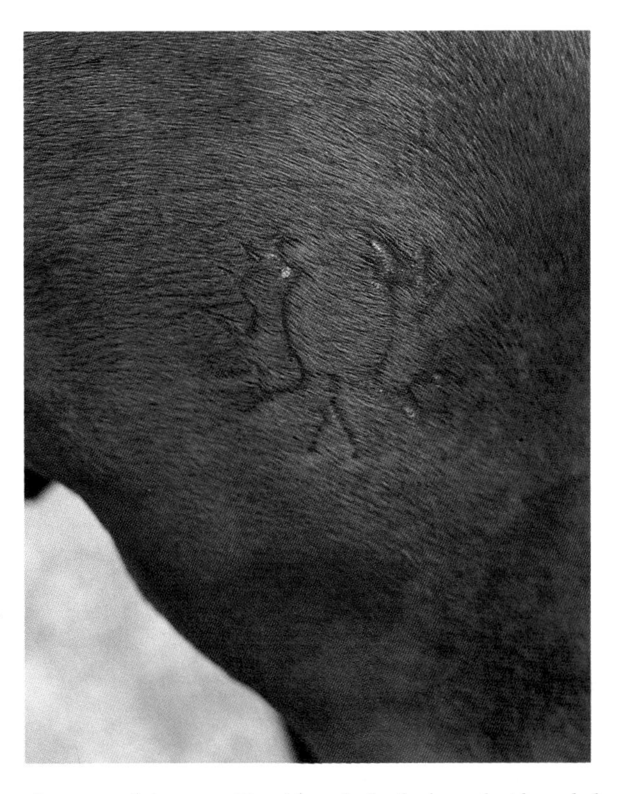

In many of the warmblood breeds the foals are hot-branded before weaning; the branding iron causes them no distress

So what is weaning all about? Your foal is now ready to be separated from his mare. His first flu and tetanus vaccination has been done and his worming programme is up to date. The question is, how do you go about this in practical terms? There are several methods, and the fact that they are all put into practice (some more than others) means that they can all be quite successful; much depends on your particular circumstances. Whichever method you choose, however, it is important that your foal is **already familiar with eating hard feed**.

Feeding prior to weaning

Begin feeding a foal about six weeks before weaning. The foal needs feed to

1 Supplement the mare's milk at weaning;
2 Familiarise him with the taste of feed;

3 Ensure that his gut becomes accustomed to different feed stuffs;
4 Make up for the fact that the grass no longer contains much goodness.

The actual type of feed that you offer the foal will depend on what you intend to feed at weaning. There are foal diets on the market, or you could feed stud diet, or traditional oats mixed with, for example, a 50/50 balancer diet. Due to the variety of types and sizes of weanlings it is beyond the scope of this book to recommend quantities of feed. However, a balanced diet with an appropriate mineral supplement is of vital importance. **Do not overfeed the yearling – being overweight can cause bone growth problems**.

On the other hand, a foal in poor condition at weaning will suffer unnecessarily, as will one who is not well. Foals who have a cough or snotty nose should not be weaned until they have recovered.

METHODS OF WEANING

Group weaning

Most people will agree that this weaning method is without doubt **the least stressful**, and as close to the ways of the wild horse that we can re-enact. In fact many Thoroughbred studs have practised this way for some years. You need to have **a minimum of three mares and foals grazing together** and preferably four or more. At this particular time of the year – normally September –

they are out at grass day and night. On the chosen day lead two of the mares (preferably good friends) to another field some way away across the farm, **out of earshot** so that neither mares nor foals can hear any calling. And **remember** (unless you have done this with a particular mare before) to lead her in a bridle – she just might not like being separated from her baby, and if you use a bridle you will have better control.

An extra person should remain in the field with the other mares and foals in case they decide that they want to follow: stand at the gate with a lunge whip, although it is doubtful that it will be needed.

Take your mares to a field with as little pasture as possible; the reason being that you want their milk to dry up and lush grass will produce more milk. **Leave their headcollars on**, as you will need to catch them twice daily to check their udders (see p128-9).

Meanwhile the weaned foals should be happily grazing with their aunts and other small friends, as they are already accustomed to doing. Two days later you can take away another mare or two, and so on continuing until you are left with just one 'aunt' and several foals – a nice sight, **and no trauma for the foals**. You must use common sense with the last mare: much depends on how independent her foal is, as to whether you risk taking her away and leaving him in the field. You

One mare looking after a number of weaned foals

may decide to wait until the foals are in at night. It is often the case that the last mare of all has a very young foal at foot which is not yet ready to be separated, so until the weather turns cold they can all stay out together – for some reason, one mare remaining gives the other foals a sense of security. An alternative way is to put a kind old barren mare in with the group, so that when the last remaining mare is taken from her foal the group still has a steadying adult influence.

Subsequent stabling of group-weaned foals: The natural progression for this type of weaning is to use a barn at night for the whole group; it is an economical system of keeping foals and in this way you will still have avoided the separating drama. However, the barn **must be safe and warm**. By *safe* I mean no protruding objects or large gaps which could create draughts. Plenty of fresh air is a good thing, and obviously a plentiful supply of fresh water. **I strongly recommend tying up at feed time**: there may well be one or two foals who need different feed rations, extra vitamins or medication. So instil discipline from the beginning and tie each foal up individually. At this age it is best for them to eat from a feed bowl at stable-door height rather than on the floor.

If you have not tied up your foal before (see p136), then do not worry, he will quickly learn, especially if as soon as he is tied up the 'meal is served'! **Remember**, *always* tie the rope on to baler twine or string *as opposed* to directly to a tie tie ring; then if he pulls back the string will break and no harm is done. **Warning**: foals just love to chew ropes, so I suggest you make up your own with baler twine – a more economical method!

Unfortunately if you are a single mare owner it is unlikely that this particular method can be put into operation; however you may well consider sending the foal to a stud which practises this method, leaving him there either for a few weeks, or the duration of the winter; the longer you leave him the better, as he will make new friends and will find it stressful to be taken away from them too soon. It should be noted that it is difficult to introduce one extra foal to a herd at a later time – in other words, when the rest of the foals have been weaned a while and are well settled into their routine. Foals, like children, tend to pick on newcomers, and they must be able to stand up for themselves.

Weaning the single foal

To wean a foal on his own is tough, and undoubtedly stressful. The only sure way to do this is to bring him into a stable, and keep the top door closed until he is settled and has no further desire to jump out – this may be 24 hours or even two to three days. **Use your common sense**, and open up the top door when you are around; if he does not attempt to jump, then all is in order. He will soon become used to your feeding routine, but it isn't much fun on his own in the dark without his mother and I strongly recommend a companion – perhaps wean two foals together, preferably the same sex, or alternatively you may have an old pony who has no hind shoes on, that could be a kind stable companion.

Once you have chosen a nice large stable, put down a deep, comfortable bed of straw and **make sure it is quite safe** with no protruding objects. Check that the top door will close and that there are **no glass windows** behind bars that your foal could rear up and strike at with his front feet – I am not being ultra-fussy, I have seen this type of accident happen.

To actually separate mare from foal I suggest leading the mare into the stable and then immediately out again, leaving the foal behind, and closing the top and bottom door immediately. Then lead the mare away, in her bridle, to a field out of earshot, preferably with another companion (see p125). It is advisable to carry out the separation process in the morning so that you have the rest of the day to observe and act, if required. *Point to note*: It is not worth weaning your foal with a companion if plans have already been made for either foal to move on, for example they may be for sale. The separation process of mare and foal is tough anyway, and for him to be taken away from his newly found mate at this early stage will be just as upsetting as taking him away from his mother.

Foals enjoy playing together – that is what they do in the wild, so endeavour to find at least one

Two foals weaned together for companionship

other foal as a companion to yours during the important second six months of his life. Companionship makes a great difference to a foal's attitude in life.

Some studs, particularly in Europe, separate colt and filly foals at the time of weaning; they are of the opinion that a colt will grow up with more masculine attributes if he does not mix with fillies, even at this early age. The separation of colt and filly foals at weaning is not always practical and is not necessarily the policy of studs in England; however, it is important to observe the behavioural patterns of weanlings. If your foal is with only one other foal, **beware the bully**. Bullies can cause harm, especially to a smaller, weaker companion. You should be able to differentiate between rough playfulness and true bullying.

Although the ideal companion for a weaned foal is another foal (the same sort of age) it may not be possible to accommodate one in your specific circumstances. Other possible companions would be: a yearling, another small youngster, a pony, a quiet, older mare or gelding, providing they have no hind shoes. The older mare often makes a kind companion, especially if she has had a foal, and she may not mind the weanling looking for the 'milk bar'. It is not suggested that these would be stable companions, merely friends to go out to graze with.

Remember the paddock is as important as the stable with regard to safety. Once you are satisfied that your foal is settled (two to three days later), then it is time to put him out to graze during the day. Put on a well-fitting headcollar and make sure he is going out to grass with the chosen companion, not alone. Lead him in and out of the field correctly (not allowing him to drag behind), and make it routine to pick out his feet when you bring him in – all good training for his future. He should be quite easy to catch; the security of the mother having been taken away, he will often turn to the human for comfort.

Weaning at grass in adjacent paddocks

I have actually witnessed this method at professional studs in Australia: it does work, and I have seen it on several occasions in the UK. Its feasibility is largely due to the **introduction of electric fences**. In both cases mares and foals were seen grazing in fields fenced with white electric fencing tape. At some point in time the foals learn to respect the wire, and are obviously quite aware of its presence. They are unlikely to attempt to jump it.

At about six months of age, or when it is thought that mare and foal are not particularly interested in each other, the mare is taken out of one paddock and put into the adjacent one. The mare is not usually bothered as she can see her foal at all times; the foal remains with his 'aunts', *and* friends and mother are still in sight – in fact they can actually nose each other over the top rail. At first the mare grazes close to the fence line, and then as her foal becomes more confident, she moves further away. If there is a problem and the foal feels insecure, he calls and mother comes trotting over to reassure him. This way of weaning is certainly not traumatic for the foal in any way. He should wear a headcollar so that he can be handled daily and become accustomed to people (if not already). I like this idea of weaning, but my advice is *not* to attempt it without the electric fence.

After about six weeks it should be quite safe to take the mare out of the adjacent field without the foal being concerned.

The gradual method

When you have little land, and your circumstances are such that, for whatever reason, you do not wish to send mare or foal away, nor can you find anywhere to take your mare out of earshot, then the gradual weaning method can and does work. The foal should be at least six months old, the reason being that the mare naturally produces less milk at this time, and should be quite disinterested in her foal (unless you have a very foal-proud mare, and if this is the case an alternative method of weaning is recommended).

Those who have watched and studied youngstock believe that the first hour of separation is the most stressful, rather like the first time a child is sent off to school. This method reflects that assumption: you put your foal into a stable and take the mare out (shutting the top door). You then take the mare away as far as possible – preferably out of earshot – and then return her after one to two hours. Repeat this process on the second day, and gradually increase the time that they are apart. Obviously while the mare is away the foal should have a feed and a plentiful supply of hay and water.

After about one week the mare's milk will have become much less and the foal will be used to his new routine. By now you should be able to put him out in a field with his friends (or with other companions). Make sure that the mare is not in an adjacent field for at least one month, or if this is not feasible, use electric fencing, as suggested in the section. Basically common sense is needed. All youngsters get used to change, and although I myself do not advocate this weaning method, it is sometimes the only way for a single mare owner.

Young foals hate bitterly cold weather and wind, so do bring them in or provide shelter in these circumstances, and *never* put a weanling out **on his own** where he can see another horse two fields away. If something frightens him his natural instinct will be to try and get to that other horse, so beware – accidents happen too often!

Whatever method you choose, remember **safety first**: better to put your foal in a stable on his own with the top door shut for 24 hours than have an accident. Vets cost money.

Remember the worming programme. It is advisable to worm your foal within a week of weaning – he is no longer dependent on the mare for any protection whatsoever, so give him the best chance. Worming is expensive, but a necessity (see Worming Table p149).

The weaned mare

The mare should not be forgotten: it is all too easy to be concerned about the foal and forget your

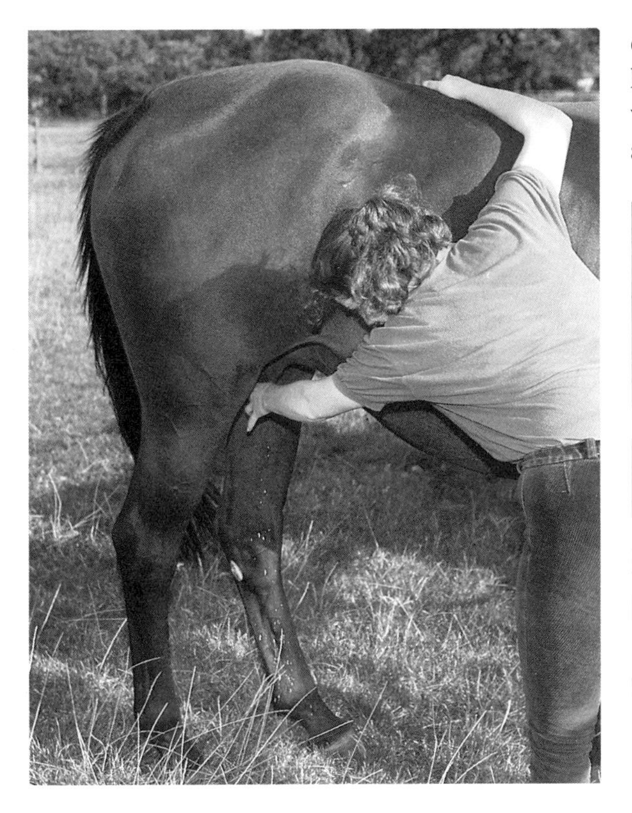

Some mares need to be milked off after they have been separated from their foals

mare four or five fields away. Check her udders twice daily for the first few days; if they are *very* full and hard, run off a few squirts from each teat. To do this you will need someone to hold her, and then, using some **cooking oil**, **vaseline** or **udder cream** on your hands, milk her like a cow. However *do not* take off much, just a few squirts, enough to relieve the pressure. **It may not be necessary to take any off at all**, especially if the foal is at least six months old; obviously the younger the foal the more milk the mare will be producing. Check her after eight to ten hours of weaning *ie* that evening and the next morning; these are the critical times. Perhaps for some reason you forget: if when you find the mare, she is walking in an uncomfortable way and the udder is *very* swollen, run off a little milk. If it appears lumpy and yellow in colour, **it is likely that she has mastitis** which is the inflammation of the udder caused by a bacterial infection. *Call your vet.* Mastitis is rare, but if it does occur the mare's mammary glands will be hard, swollen and sore. The swelling normally spreads, and as already described, when the milk is run off (if the mare allows this) it will probably be lumpy in consistency with the look of curdled milk. **The mare needs antibiotics and some local treatment**.

SUMMARY: WEANING

1. Check that the foal's first – if not his second – flu and tetanus vaccination has been done at least three weeks prior to weaning (see p112).
2. Make a decision on which weaning method you wish to adopt.
3. Make sure that the foal is accustomed to eating hard feed and is in good physical condition.
4. Ensure that the foal is free from any cough or snotty nose.
5. Wean in a safe environment *ie* check carefully the relevant stable or barn and paddocks.
6. Remember your mare after weaning. Check her udders twice daily for two days, and then daily thereafter for at least a week.
7. **Worm the foal one week after weaning.**

When can you ride your mare?

Begin to think about working your mare approximately four weeks after her foal has been taken away, preferably not sooner. Give her time to adjust, as weaning can sometimes be stressful for the mare, too. Her udders should be quite small by this time (although the bags will still be visible). Remember, she has probably not been worked for about a year or more, so take it easy. Put her into work as though she were a hunter coming up from grass – her riding muscle structure will have deteriorated even if she looks good, and also remember she has not had a girth done up for a long time.

If you are bringing a youngster back into work *eg* a four-year-old who was lightly backed and then put in foal, I suggest you begin the breaking process all over again. Having bred a foal, a mare will have matured mentally and physically and should adapt quickly to her new lifestyle.

Some owners have a wish to ride their mare sooner, perhaps when the foal is still at foot. Physically there is no reason why this should not be done; in Germany I have visited a stud which liked to earn money from its brood mares whilst they were suckling a foal – it would invite the locals to come and ride the mares in an outdoor arena, and the foals would just trot around with them. No doubt an exercise enjoyed by all concerned, and no harm done!

Travelling the mare and foal at weaning

1 *Mare is going away to another location:* In this case, when your lorry or trailer is ready to leave, put boots on the mare and take mare and foal into the pre-prepared stable (see p126). Leave the foal in the stable with the top door closed, and take your mare away from the stable straight to the vehicle and up the ramp. Do not hang around, set off immediately.

2 *Foal is going away to another location:* I suggest travelling the mare and foal together if the location is not too far away and your mare boxes well (see Loading practice p118). Once the foal is in his new location, re-load the mare as described previously.

3 *Taking mare and foal to stud for weaning:* This should be quite simple and **non-stressful**. You will need to leave the mare at the stud for about a week to ten days (or as recommended by the stud). She will be put out to graze with other mares and foals (see Group weaning method p124). After the foal has met and come to know his new friends, the mare will be taken away well out of earshot. She will be attended to as deemed appropriate by the stud, and you should be able to collect her in about a week. The foal would be able to leave about four to six weeks later, should you decide not to leave him there for the winter.

4 *Travelling a foal alone:* I do not recommend travelling a foal as soon as he is taken from his mother. If for some reason you cannot travel the mare, or the foal has been sold, I suggest you separate him from his mother for at least three days before the travel date. However, if you use a lorry with well-padded partitions, and the foal has already been taught to tie up, then you may be alright, especially if, for example, a pony or mare companion travels alongside. Un-weaned foals are difficult to load without their mother, so I suggest you lead the mare up into the lorry whilst the foal is led up following. Close the partition and unload the mare whilst someone remains with the foal; begin the journey as soon as possible. Travel a weaned foal in a similar way – in this case it may help to lead a pony up in front to give the foal the incentive to go up the ramp. If you have difficulty in loading it is not wise to use a lunge line behind a foal; I suggest you have someone at his head and one strong person either side of his hindquarters. If he is small enough they can join hands behind his quarters and gently push him up from behind. With a larger foal a wide, thick piece of material, held behind and on either side by two people, may assist in firmly pushing him up the ramp.

5 *Travelling a foal alone in a trailer:* There are two different ways of travelling a foal in a trailer, and which one you choose will depend on the prior handling of the foal. If he has been well handled, is easy to lead and will tie up, I suggest you travel him tied up, as you would an adult horse, with the central partition in position. Put a bale of straw under the chest bar to prevent him from trying to escape under the bar, making sure that the baler twine is not evident. I suggest you start the journey by travelling with the foal to make sure that he is not frightened, and have a bucket of feed to offer him if necessary.

The alternative way to travel your weaned foal is to take out the central partition, put in plenty of straw banked up at the sides, and travel the foal loose with the top doors closed and some hay on the floor. Make sure that the small side door is locked or securely closed so that if the foal leans on the handle the door does not fly open during the journey.

PURCHASING A WEANLING

Why buy a foal and not a ready-made horse?

1 To buy a weanling is, without doubt, a **moderate capital outlay** in comparison to a ready-made horse.

2 You can have 'the pick of the bunch' from the stud's herd.

3 You have the often rewarding period of bringing up a youngster from weaning.

4 You can follow your own discipline and training methods.

When purchasing a weanling, have a good look at him away from other foals

But are these reasons justification for the three subsequent years of patience, worry, ailments and expenditure, particularly as there is no actual guarantee that the 'finished article' will turn out as you originally envisaged? Would it not be wiser to buy a ready-made horse in that at least you would be able to sit on top and 'try him out' before handing over the money?

'To buy or not to buy' a foal – that is the question. Much will depend on your personal financial circumstances. You may well be able to afford the price of a top quality foal but not that of a top horse – the foal you buy may turn into something you could never afford as a three-year-old. Breeders need to sell, but they are reluctant to do so once they see that their foal has matured into a two-year-old with obvious potential as a riding horse; by then it is therefore usually too late for *you* to purchase.

So you have decided, and would prefer to buy now and enjoy the next three years. Have you somewhere safe to keep your new purchase *ie* a suitable environment for a young foal? Will he have at least one other foal as a companion? Are you prepared to spend time on educating your youngster? When the time arrives to back him, are you capable of doing so, or are you prepared to send him away to be professionally schooled? These are all questions that need to be considered before you even start searching for your ideal foal.

Finding the right foal for your needs

Having made a positive decision to purchase a foal, you need to decide what type of horse or pony you are looking for: what do you want him

(Above) *A weanling with poor conformation*

(Opposite above) *Study the mare's conformation and pedigree before purchasing a youngster*

(Opposite below) *Ask to look at a photo of the sire, and consider his credentials carefully*

to do in later life – do you want potentially an all-round riding horse, a family pony, a dressage horse, show jumper, eventer, or maybe for breeding stock? Are you looking for a specific breed?

Whatever line you take, in addition to searching the 'Horses for Sale' sections of journals, remember to contact the relevant breed societies, and also breeders. They are likely to know what youngsters are on the market, and will nearly always be happy to supply you with names and addresses.

What to look for on viewing

First impressions are all-important: you must first like what you see. But then look further:

1 Inspect conformation – has the foal a large, kind eye? Has he presence?

2 Look at his walk and trot, in a straight line on hard ground.

3 Watch the foal running loose, his paces and his reactions.

4 Look at the mare (if the foal has been weaned and is not together with his mother).

5 Go and look at the sire if he is nearby.

6 Request to look at any other offspring from that particular mare. Have they performed?

7 Study the pedigree and bloodlines.

8 Check that all his registration papers are in order – contact the relevant breed society if in any doubt.

9 Be satisfied with the way that he has been brought up until now, *ie* stable management, worming procedures and so on.

A well mannered weanling being shown to a potential customer. Viewing a foal that has been handled makes the purchase decision much easier

Veterinary examination

Having found the foal that you would like to buy, I recommend that you have him examined by a veterinary surgeon. It is normal to use a vet that is not the regular one attending the yard from where you purchase, but sometimes this is unavoidable. The examination cannot be as complete as that for an adult horse, but nevertheless, the basics will be checked: eyes, heart, lungs, mouth, limbs, overall soundness and conformation. I advise that you are present when this takes place so that you can ask the vet any questions that may arise.

When to take your new foal home

I suggest that you do not consider taking the foal until he has been weaned. It will make for a much safer journey if weaning is well over, and the foal has been handled enough so that he will load without too much trouble. Check when he was last wormed and make sure you are given the registration papers and vaccination certificate (once the money has changed hands). Ask for details of the type and quantity of feed that he is being given and prepare for his homecoming (see p130, Travelling a foal alone).

After all these decisions I hope that finally you make the right one – have a bottle ready to celebrate!

Your foal in his new home

It will take a while for your newly purchased foal to become accustomed to his new environment. Not only has he been taken from his mother, but all that was familiar to him has disappeared. He will need time to adjust. If the companion that he will be going out to graze with is an old pony or mare then he should be quite safe, as these will most likely get on with their grazing and not be unduly disturbed by the new youngster. However, if the companion is to be another foal I suggest you introduce them over a wall or stable door and then put them out in a *small* paddock, which will curtail the otherwise inevitable and endless galloping around a large acreage. If he is to go out with several foals, allow him to meet just *one* first, for a day or so, so that they make friends, before putting him out with the entire herd.

When to castrate (geld) your colt foal

Colt foals may be castrated prior to weaning at four or five months old, although this custom is not practised very often today as it is thought that the foal should be given more time to mature and have no setbacks at this early age. Others maintain that gelding is such a minor operation when a foal is only five months old that there would be no trauma; the mother and her milk will provide comfort for any pain. Weight loss is considerably less if the foal has already been accustomed to eating hard feed (see p123).

The most usual time for castrating (gelding) a young colt is in the spring of his first year (see p141).

A home-made foal muzzle: a flower-pot with large air-holes drilled through the bottom and a fillet string for the head-piece. This is used to prevent the foal taking any milk (for example for eighteen hours before anaesthetising for castration)

AFTER WEANING: THE WINTER MONTHS

Once the evenings become cold, and if there is a lot of rain, then it is time to bring your weanling in at night. Set yourself a routine and stick to it. Weanlings do best when they are out for a few hours during the day, and closed in at night; this ensures that they rest – if they merely have a field shelter (without doors) they will not settle down and rest properly, but will be continually in and out in response to Nature's wild night-life.

The method you used to wean your youngster will dictate whether it will be stabled alone, with a companion, or in a group situation using a barn. The latter is certainly the simplest, most labour-saving method and allows plenty of room for the young horses to move about; should weather conditions be very poor, there may be days when you cannot put them out to graze.

Feeding should be done individually. If you have a number of yearlings I suggest that you tie them up to feed them; inevitably there will be one less bold than the others which will not receive his correct ration of feed, because even at this early stage in life a 'pecking order' will be prevalent.

I advise tying up each youngster adjacent to its own feed manger. This is also an ideal way to teach a youngster to tie up; it is free of any trauma, and the yearlings become so accustomed to the routine that when they come in from the field they will automatically go to their own manger and wait to be tied up. Their feed completed they will then be quite content to stand for a while before being untied – though beware, once the feed is consumed they tend to move on and eat the rope (so it proves less costly to make your own!).

Some of the hardier native breeds may well remain out all year round, but even they will require regular feed, vitamins and hay.

Feeding the weanling

Every owner and breeder has his own ideas on feeding. Your youngster needs to be fed according to his type, size, height, bone and actual growth rate (see chart p140) – and remember, your eye is a good guide, too. During your youngster's first winter he should not become overweight – you should just be able to feel his ribs, and there should not be any excess fat.

What type of feed to use is a matter of opinion. Some studs today are moving over to modern prepared feedstuffs for defined age groups; others continue to use traditional oats and good clean hay. Whatever you decide to feed, avoid bran, which is proven to be detrimental to good bone growth, and remember to add an appropriate vitamin supplement containing the correct calcium to phosphorus ratios (see Physitis below). Seek advice, but don't talk to *too* many folks as you will get in a muddle!

Physitis (Epiphysitis)

This term describes abnormal activity in a growth plate, and normally occurs in fast-growing weanlings and yearlings. It usually affects the lower growth plate of the radius, and can be seen as a slight swelling just above or around the knee. Sometimes swellings can be seen around the fetlock joints. The youngster will not necessarily be lame, but when pressure is applied it may cause pain. Discuss the situation with your vet; he will probably suggest a restricted diet combined with restricted exercise and supplementary minerals. The earlier you call the vet, the more chance there is that with correct management, the swellings will subside (see also chapter 6 p110).

Re-location during the winter months

Your weanling is now no longer reliant on his mother; he is content with his new friend or friends, and hopefully he may remain with them for the duration of winter. However, if for some reason you do have to move him to a new location, be prepared for problems. Remember that at this stage in his life (about 7–8 months) he is still quite insecure, and has in fact transferred his dependency from his mother to his new companions; so use your common sense if you have to move him: ensure that any young companion left behind remains with a friend, or is stabled – and if he is not accustomed to being stabled alone he may still attempt to jump out. Arrange for your weanling to travel supervised, and preferably alongside another horse or pony. Moving a youngster at this stage is similar to weaning a second time.

Vices

Vices as we know them in the older horse, often originate from his days as a youngster. Avoid stressful circumstances; for example:

1 Do not leave a weanling in a barn on his own; if he has to be kept in for a few days, make sure that he always stays with a companion.
2 If a weanling is stabled together with another at night, and one has an injury meaning that it has to stay in during the day, make sure that it can cope in the stable on its own. If it becomes very bored it may start sucking the door, or the manger, or its bucket – the beginnings of wind-sucking.

3 If you put a weanling out in the paddock on his own, ensure that he is supervised – something might frighten him and he could try to jump out. Alternatively he may be bored and start to nibble the fence-line, and this could be the start of crib biting.

Handling

If your weanling is put out in the field every day, then it is likely that the amount of handling this involves is sufficient, providing you pick out his feet regularly. However, if he is simply run out daily (as opposed to being led), I advise a specific handling session about once a week to ensure that he retains good manners and remembers what he learnt as a foal alongside his mother.

SUMMARY: MANAGEMENT OF WEANLING DURING THE WINTER MONTHS

1. Make sure that your weanling has at least one companion, preferably of a similar age.
2. Make sure that there is a suitable barn or stable for the winter nights.
3. Feed according to size, type and condition.
4. Add an appropriate vitamin supplement with a sufficient calcium to phosphorus ratio to maintain correct bone growth.
5. Worm every four weeks.
6. Arrange regular visits from the farrier.
7. Do not subject your weanling to stressful situations which may lead to bad habits, and potential vices.

8

YOUNGSTERS: THE FIRST AND SECOND YEARS

During his first year a foal will develop rapidly and will demonstrate varied patterns of behaviour; colts will probably be gelded. For the twelve months following a foal's first birthday he is known as a yearling; note that 1 January is the official birthday of the Thoroughbred horse, whatever the date he was actually born on. Throughout the first two years of a young horse's life he should be well handled and taught good manners. He should receive routine care from the farrier, be wormed and vaccinated regularly, and enjoy the benefit of sound feeding methods; daily observation and plain common sense are prerequisites to be expected from any owner and if the owner does not have the time to attend to the yearling himself, then he should delegate the responsibility to someone who is sufficiently competent. Your youngster is likely to go through some most unattractive stages in looks and manners, but you must be patient – he will probably change for the better!

This chapter will discuss how much to handle a youngster – too much can lead to as many difficulties as too little; also, the important decision of when (or if) to castrate the young colt is explored and followed by an explanation of what castration entails; and we explore the trials and tribulations of keeping an entire without adequate facilities. Finally the benefits of in-hand showing are considered and advice given on preparing the youngster for the show ring.

EARLY GROWTH AND MANAGEMENT

When should a yearling be turned out to grass? Certainly during the winter months keep to the routine you have followed since you brought him in the previous autumn – that is, he should stay in at night until the spring and only turned out completely when the nights are warm and dry. Do *not* be tempted to leave him out whilst there is still a likelihood of cold wet rain and before the spring grass is well underway – he will only lose condition, and all the good feeding programme that you have followed during the winter will have been wasted. As the weather improves leave him out for longer hours, reducing his feed over a couple of weeks as increasingly he benefits from the rich spring grass; ultimately he will be out for the whole twenty-four hours with no extra feed.

Plan ahead: arrange for the blacksmith to trim his feet before turning him away, and preferably worm him at the same time. A few weeks of freedom from all handling will be of great benefit: therefore look at him daily but **leave him in peace**.

Growth patterns

It is interesting to note that:
1 at birth a foal's wither height is around 60 per cent of its adult height;
2 approximately 28 per cent of a horse's total growth in height is attained during the first year;
3 the bodyweight at birth is around 10 per cent of the final weight as an adult horse, and develops about 50 per cent during the first year.

These growth factors are shown on the following charts, and demonstrate how important it is that a youngster is given the maximum opportunity for growth during his first year through good feeding and routine management.

The measurements given on the following charts are taken from a Thoroughbred; however most breeds of horses and ponies follow the pattern shown.

Growth is not greatly influenced by environ-

Growth chart 1: *Growth curves showing the* height of the withers *and* canon bone development *during the first twelve months. Note that as a foal the filly grows in height a little faster than the colt, and then in the eleventh month the colt overtakes the filly.*

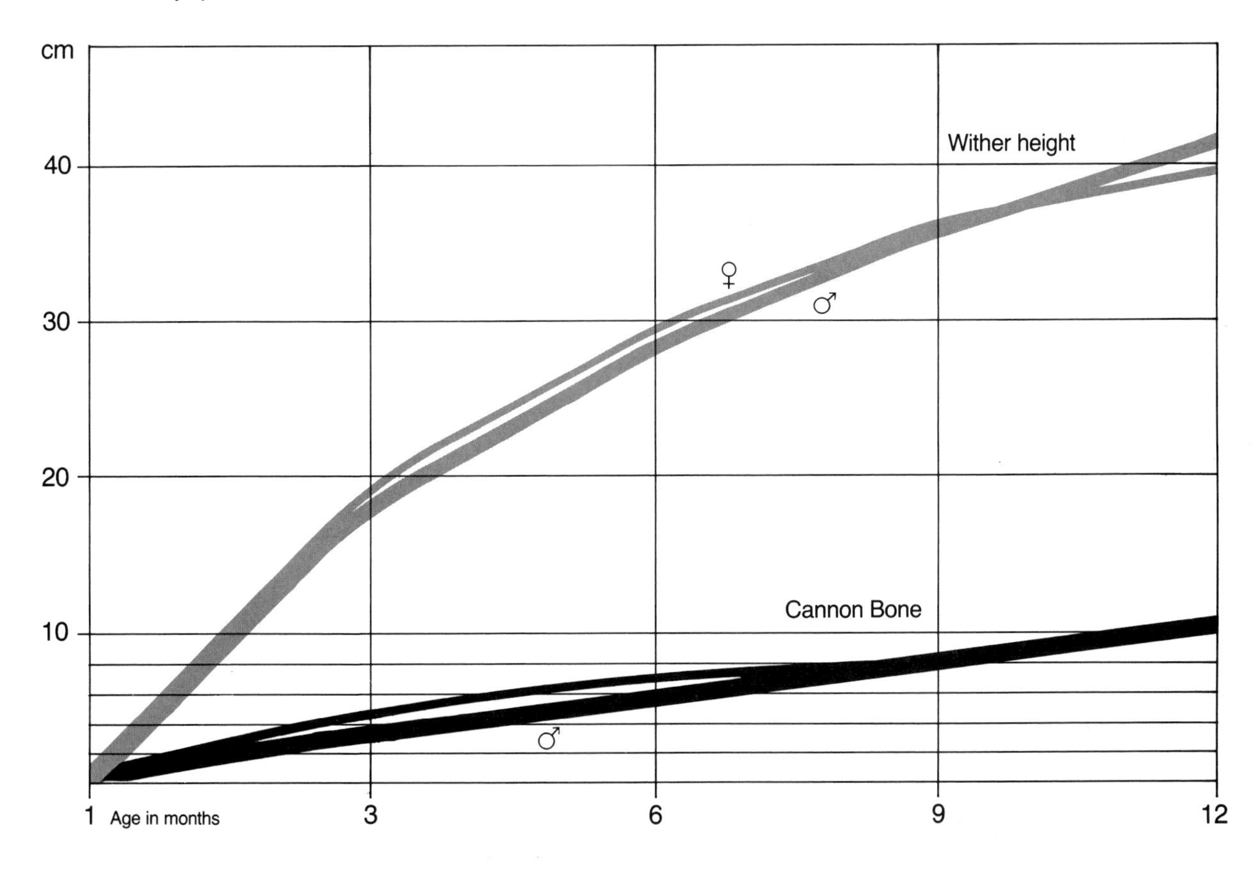

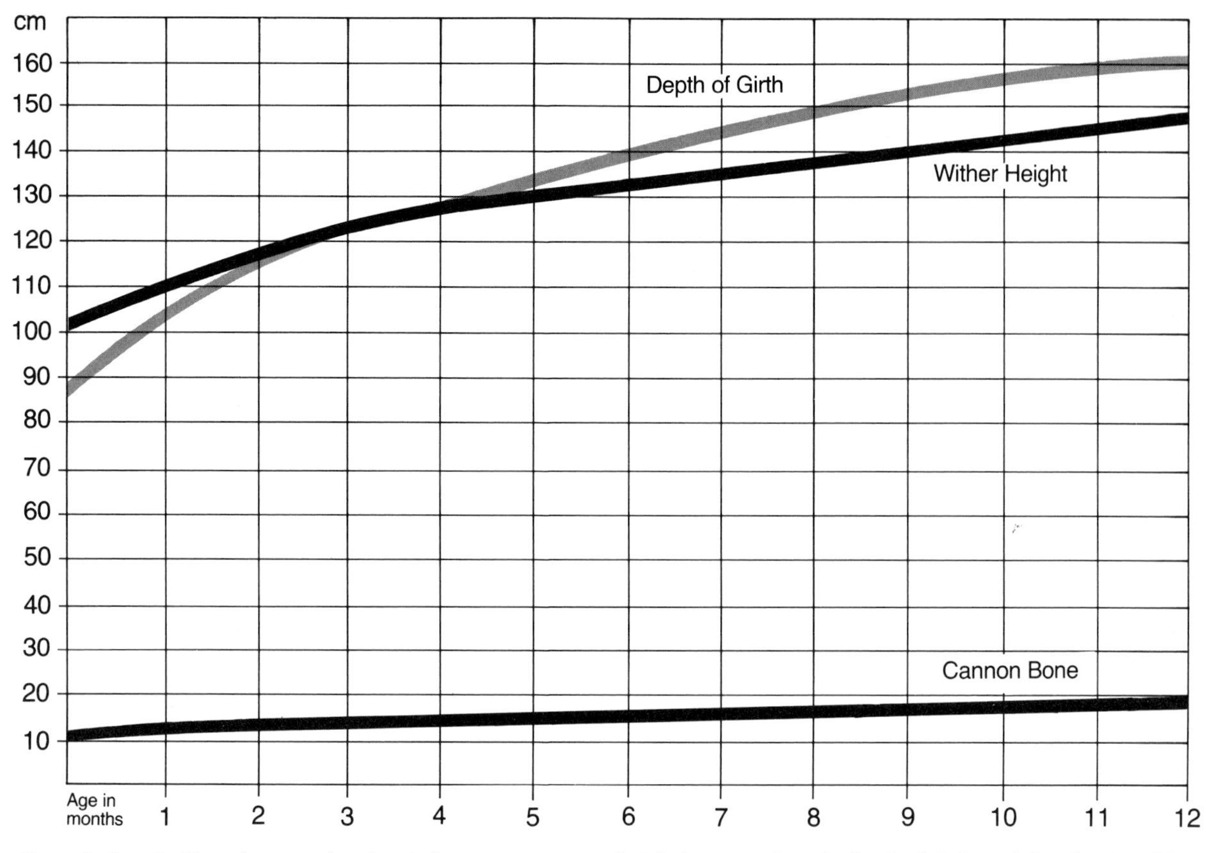

Growth chart 2: *Growth curves showing* girth measurement, wither height *and* canon bone development *during the first twelve months. Note that until the fourth month the wither height is greater than the depth of girth, and then the size of the girth overtakes the wither height allowing more space for the growth of the heart and lungs*

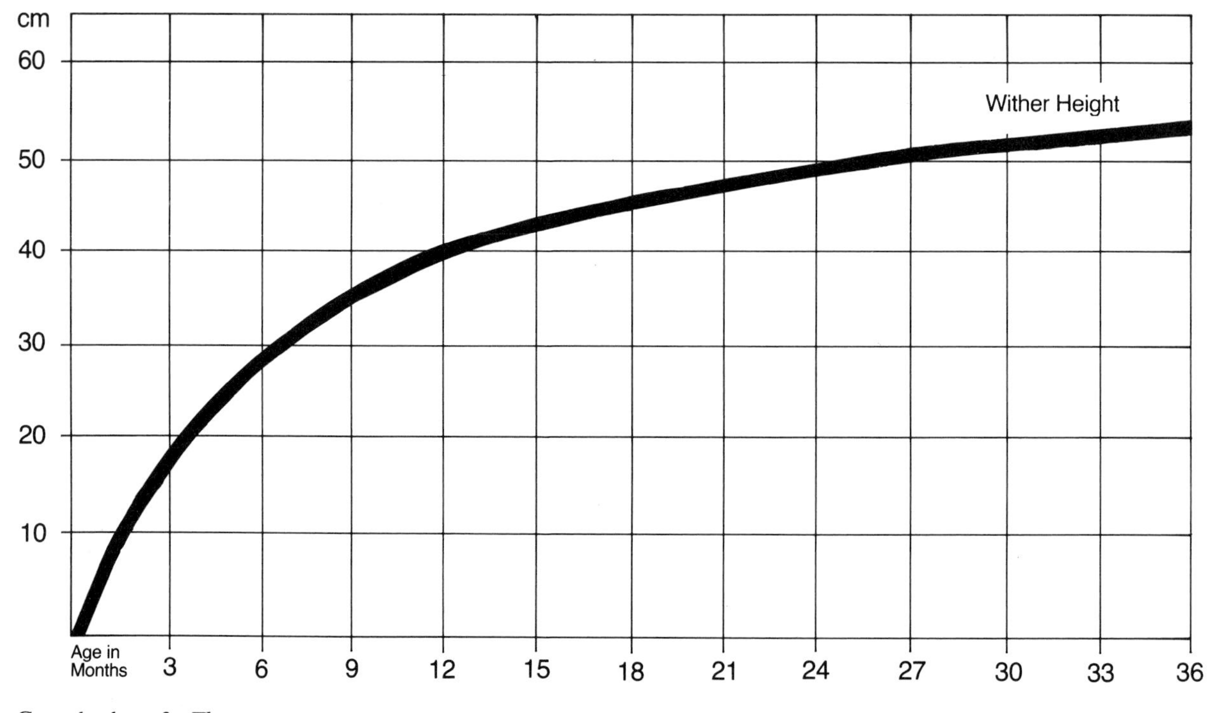

Growth chart 3: *The average increase in wither height of Thoroughbred horses of both sexes from birth to 3 years old*

mental conditions, except when a youngster is subjected to stressful situations, and as a result of sickness.

Castration

The operation known as castration or gelding involves the removal of the two testicles; this renders the horse sterile, and consequently more docile and easier to handle since there is no further sexual attraction towards mares after a certain period.

At what age should the colt be castrated? A colt may run with other foals for some time, but it is possible that his sperm is fertile in the spring of his first year, although it is unlikely that a filly of the same age would conceive. None the less, colts and fillies should be separated in the spring when they become yearlings; at this time colts are likely to become a nuisance to fillies and are better kept with other colts or geldings. In large studs colts and fillies are normally separated at weaning. Most owners of colt foals arrange to have them gelded in the spring of their first year; at this age there is rarely a problem with resultant swelling, and because they are put straight out on good spring grass they have the optimum chance for continued growth. Once a colt is castrated, any behavioural stallion characteristics disappear very quickly; physically, a crest will not appear and the neigh will become higher pitched. It should be noted, however, that **the sperm in the glands of a colt can remain fertile up to approximately six weeks after the operation**.

The operation can be carried out with the horse standing or with him lying down. If the latter method is chosen, he may have to be taken to a veterinary operating theatre and in that case should have no feed or water for the previous twenty-four hours. The operation can be done at your own premises, providing your vet is in agreement; however, some vets are not keen to operate away from their theatre in case complications arise, and there is added risk for the operator, especially with an older horse, if he carries

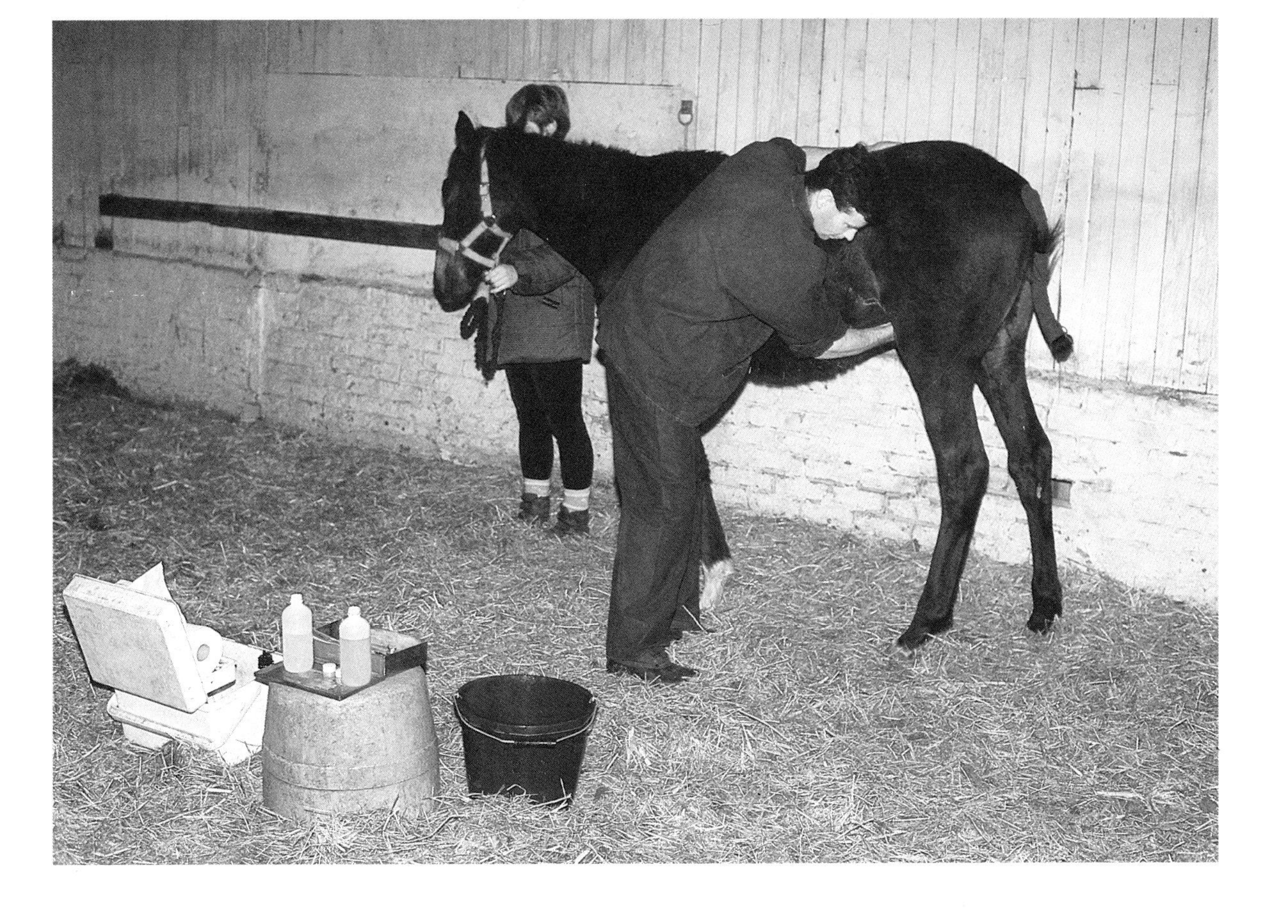

A yearling being prepared for castration at home

out surgery whilst the colt remains standing.

The operation itself only takes about fifteen minutes. If carried out at your own premises the colt will be given a local anaesthetic and may go back out to grass immediately after surgery. However, if you take him to the vet's operating theatre, be prepared to leave him there for at least half a day as he will need a period of two to three hours to recover.

When to castrate: The operation of castrating can be carried out at any time, but it is advisable to have it done in the spring or autumn, for two reasons:

1 After the operation the gelding needs to be constantly on the move to help keep down any resultant swelling; it is recommended therefore that he should be out at grass day *and* night. If there is no permanent grazing available he will

(Left) *Castration taking place whilst the sedated colt remains in the standing position;* (below) *One testicle has been removed;* (right) *Colt on the operating table during castration*

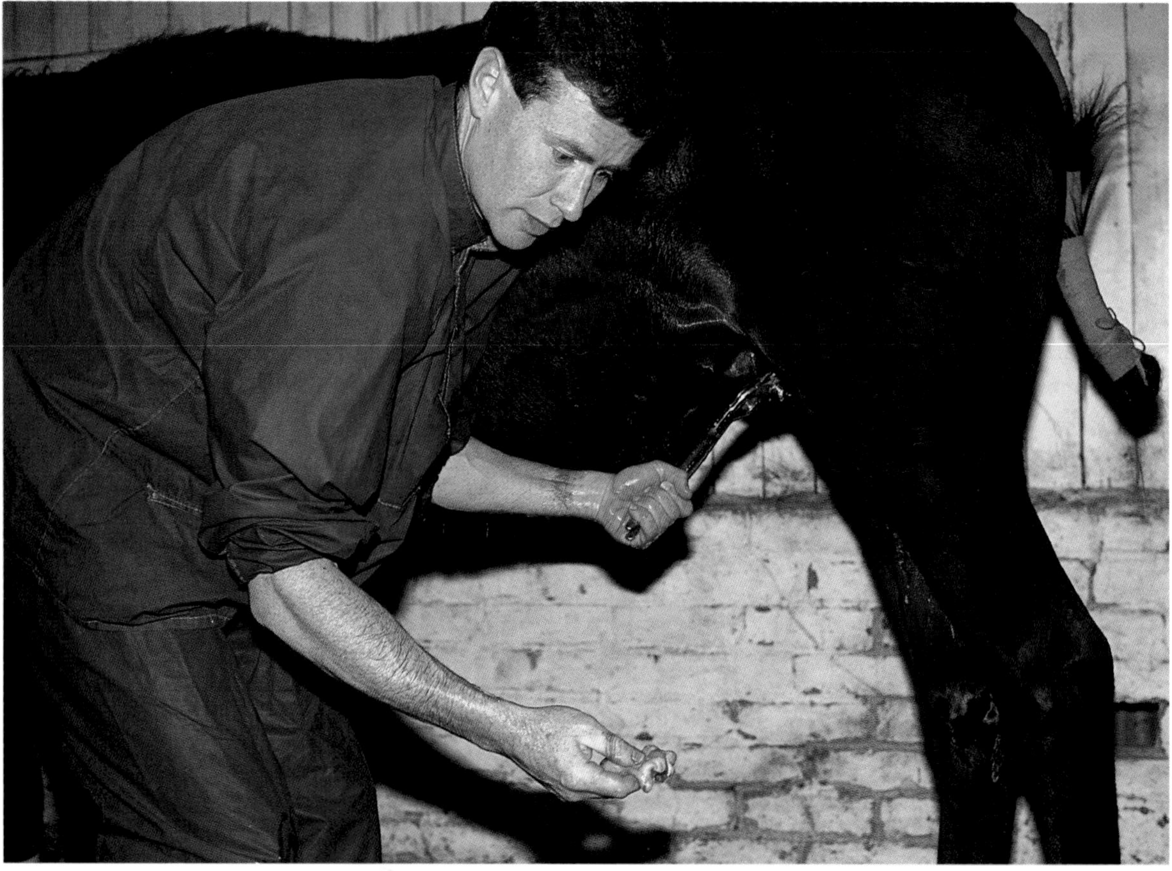

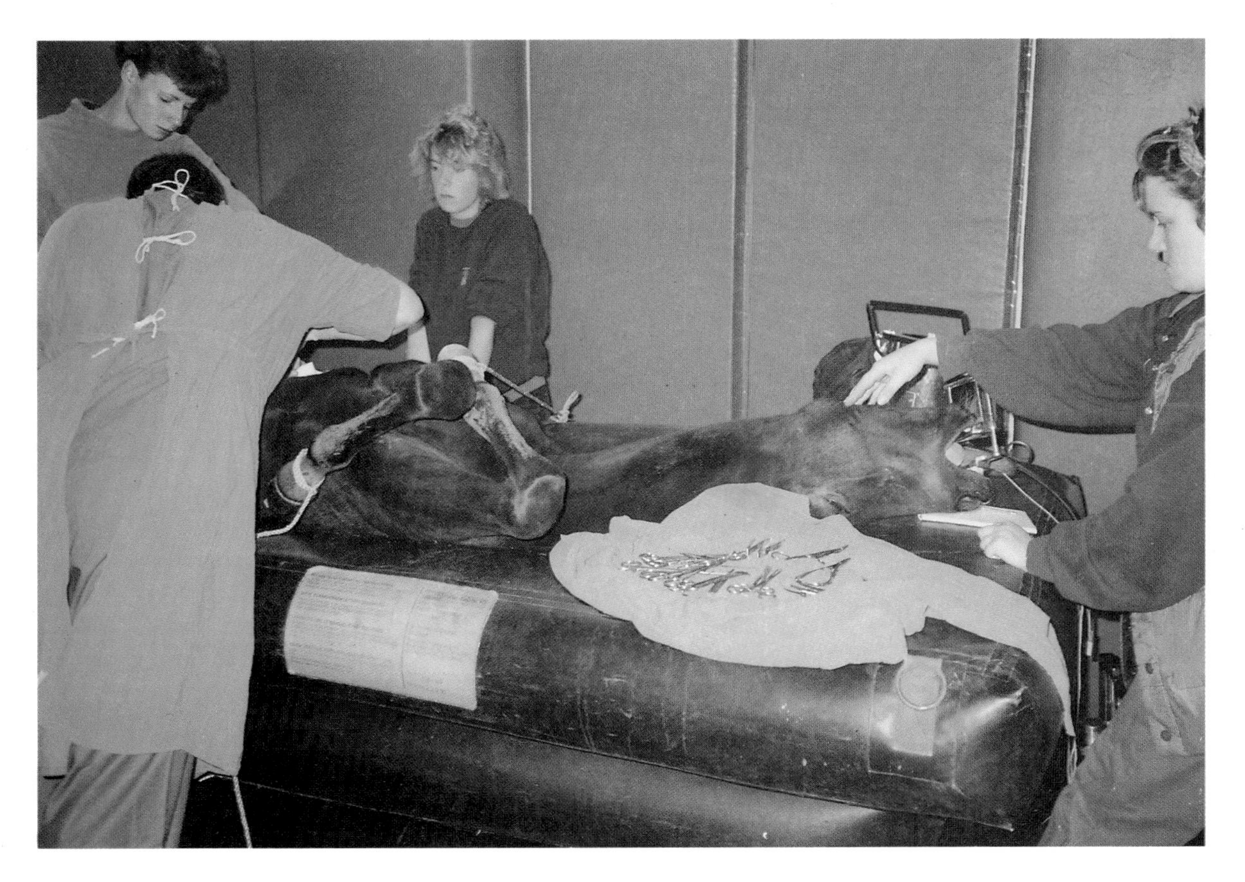

need to be walked or even lunged every two hours for the first few days until the swelling has subsided.

2 The weather conditions should be temperate, but with no flies which would settle on the open wound resulting in infection and further swelling.

Keeping a colt entire: Today, more competition riders, especially those in the higher echelons of dressage competition, prefer to ride stallions. A stallion is likely to have more presence than a gelding, he will have a more strongly developed muscle structure, and may well be more intelligent to work with. However, if you do decide not to have your colt gelded, bear in mind that bringing up an entire is not easy unless you have suitable facilities and an experienced handler.

If your young colt shows outstanding qualities and movement as well as having a good pedigree he could possibly develop into a breeding stallion. If this is the case he should be given the chance to procreate, and be kept entire until such time as he can go forward for stallion approval by the relevant breed society.

POINTS TO NOTE IF YOU INTEND TO KEEP A COLT ENTIRE

1. By the spring of a colt's first year he will show stallion characteristics, and should be turned out with other colts or geldings.
2. As a two-year-old he will probably have to learn to live on his own, as it may be difficult to find him a companion he will accept: he will become increasingly playful towards other male horses, and if put to graze in a paddock adjacent to mares or fillies he will almost certainly try to break through to them.
3. He will need to be handled regularly by a competent, experienced person.
4. He should lead as normal a life as possible, and if he is not happy with his day-to-day routine he will be likely to acquire bad habits and vices. Naturally the older he becomes the greater his masculinity will be.

Castrating the older horse: Because stallion characteristics are fashionable today, some breeders who do have suitable facilities are leaving their youngsters entire until they are three or four years old and then having them castrated.

Stallions can be difficult to manage; if you intend to keep a colt entire, you must have suitable facilities and a competent handler

(Opposite above) *Keep the youngster happy so that he does not learn bad habits such as crib-biting*

(Opposite below) *Swollen scrotum nine days after castration on a three-year-old; colts should not be allowed to remain idle after the operation*

SUMMARY: CASTRATION

1. Decide when you want your colt castrated.
2. Discuss with your vet whether he will carry out the operation at your premises, or in an operating theatre.
3. If castration is to be done in the veterinary surgery, remember a) to practise loading a few days before, particularly if your colt has never travelled alone (see p130), and b) to starve the colt of food and water for the 24 hours before he is due to be operated on.
4. Ensure that you choose a time of year when it is warm enough for him to be out at grass 24 hours a day, and when there is little evidence of flies; and make sure that before the operation he is used to being out at grass night and day.
5. Remember that his sperm may remain fertile up to six weeks after the operation.
6. An older horse may retain stallion behavioural characteristics for up to four to six months after the operation.

However, not all young colts develop a crest or a particularly masculine muscle structure until much later on, perhaps not even until they have covered mares.

In older horses the recovery period after castration takes longer; the testicles tend to swell more, and the behavioural patterns typical of a stallion may take up to four to six months to change due to hormonal activity. Moreover if a stallion has served mares, he may always show stallion characteristics in the spring, that is, in the natural mating season.

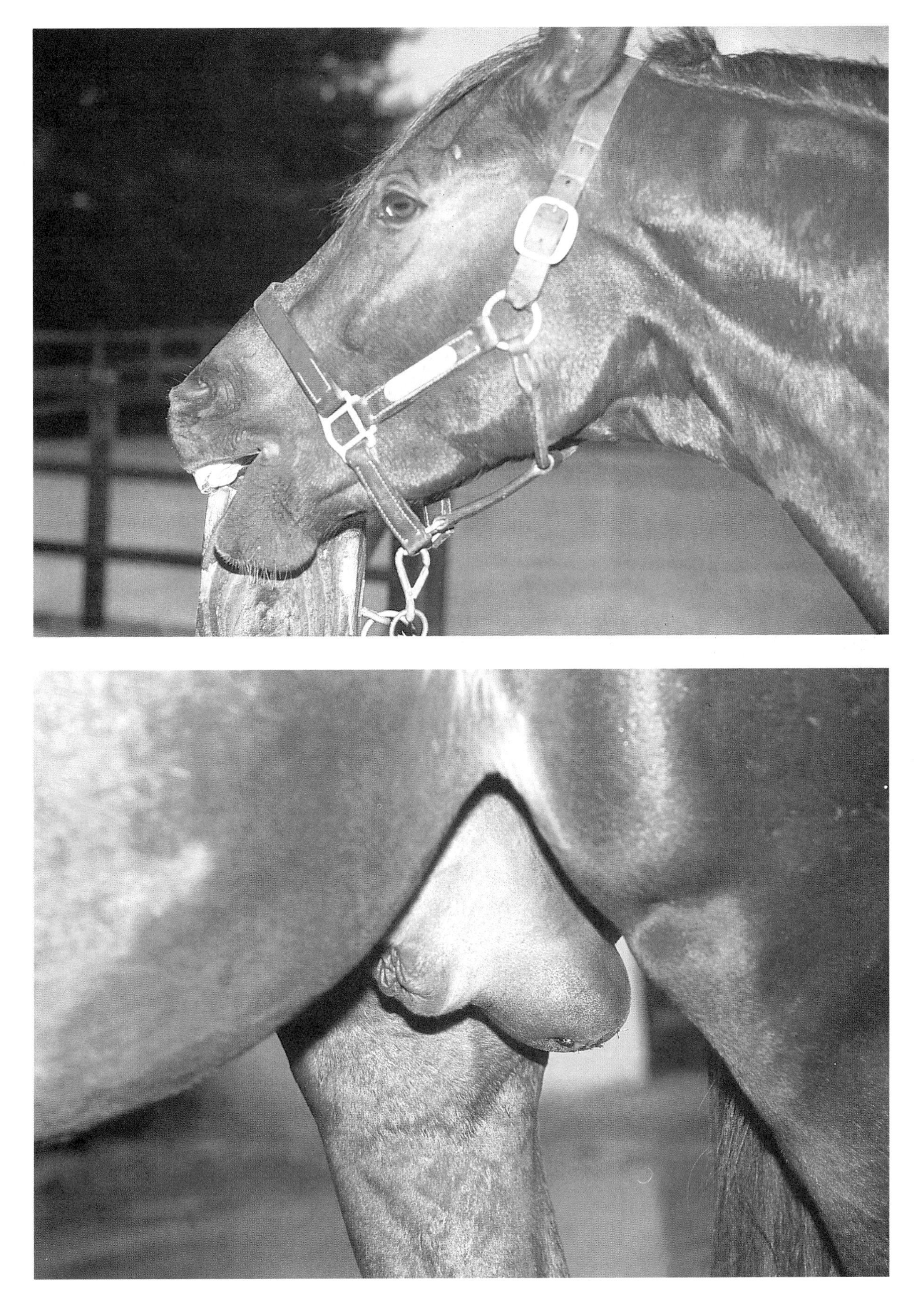

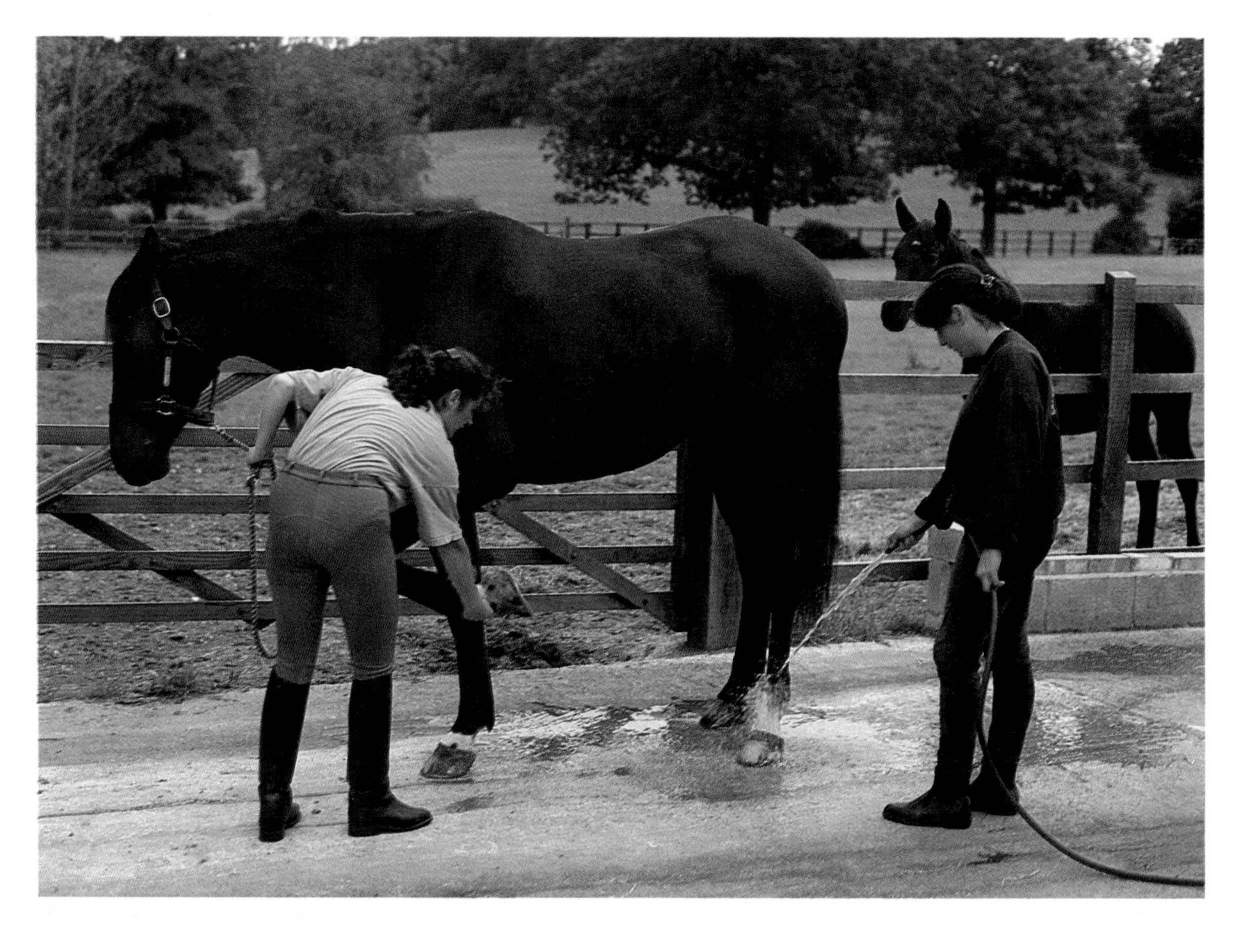

Teaching the youngster to accept water on his legs direct from the hosepipe

(Below) *Cracked heels are often seen on youngsters, especially those with white socks. Treat on a daily basis with a proprietary ointment (such as zinc and castor oil)*

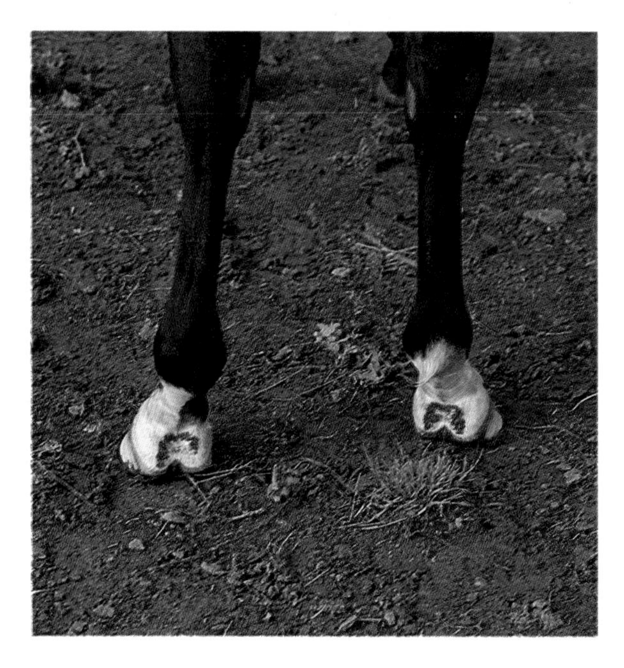

THE DEVELOPING YOUNGSTER

Yearlings and two-year-olds go through interesting stages – some depressing and others quite pleasing – and there is much change in growth and behavioural patterns. So if, for example, your yearling looks very plain in the head, do not despair, as it will probably improve as he develops and matures. If he turns in a toe slightly, don't panic, simply make sure that this fact is brought to the attention of your blacksmith who may well be able to improve the problem, as long as regular, frequent visits are arranged.

Splints

The term 'splint' is used to identify a bony enlargement in the region of the canon bone.

Although splints can appear at any time during a horse's life, they usually occur between two and four years old in either the front or hind limbs, but more commonly in the forelegs.

A splint may develop in a young horse for a

number of reasons:

1 Following a knock or blow on the leg, which could be self-inflicted.

2 As a result of a small tear to the ligament between the splint bone and the canon bone.

3 As a result of incorrect proportions of phosphorus and calcium in the diet.

4 As a result of poor leg conformation or incorrect trimming of the foot, which made the youngster place his foot down unevenly on the ground.

The young horse is not always unsound during the formation of a splint, but is most likely to be sound at the walk and unlevel at the trot. During the first stage of formation the swelling is normally soft, and any pressure may cause pain. This swelling is in part bony and in part overlying fibrous tissue. Consult your vet if lameness persists; he may recommend a period of box rest. In the latter stage of formation the swelling will become firm, and when all inflammation has ceased the horse will be sound.

The bony swelling can be removed if required for cosmetic reasons, but the outcome is not always successful.

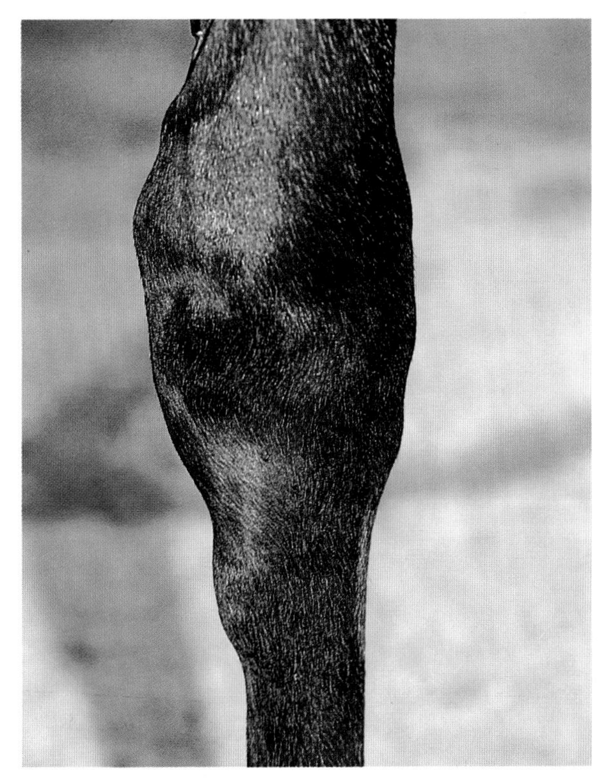

A fully formed splint below the knee

Swollen Lymph Glands

Youngsters sometimes have swollen glands around the angle of the jaw. These are an indication that **infection is present in the area**. Treat it as a warning. If the youngster shows any signs of listlessness, runs a temperature or has a snotty nose, **do not hesitate to call the vet.**

Providing he is well in himself the swellings, given time, should go down – they could be the result of a summer cold, or an allergy to some foliage. **But if the symptoms persist or the swellings increase in size, seek advice from your vet.**

Soft lumps which appear above the chin groove could be the swellings associated with a new tooth coming through. Providing the youngster is not in pain, keep a watchful eye on him, but do nothing.

Teeth

The teeth of a foal through to two years old do not normally cause any problems. However, at some time between two and four years the temporary incisors are shed and the permanent incisors appear abruptly through the gums, and during this period the mouth should be checked regularly. There are very identifiable growth patterns in teeth, especially in the young horse, and people have estimated a horse's age by looking in his mouth for centuries.

Wolf teeth: A wolf tooth is a vestige of a more primitive ancestor, and has no function in the modern horse. There can be up to four wolf teeth present in the mouth, one in front of each first molar, but it is more usual to find them in the upper jaw only.

A wolf tooth is normally small, with short roots and a sharp pointed end, which is likely to cause discomfort when the horse is bitted. At this stage they should be removed, and your vet can normally do this quite easily without a general anaesthetic (see also p148).

Warts

If your yearling suddenly appears with warts all over his nose, don't worry: these are usually the

APPROXIMATE GUIDE TO THE YOUNGSTER'S TEETH

Between one and ten days: two central incisors present, plus three cheek teeth.
Between four to six weeks: the lateral incisors will appear.
Between six and nine months: the corner incisors will appear. These are temporary teeth, relatively soft, almost white in colour and come to a point at the base.
At one year: all six temporary incisors are present, plus the three premolars and the beginning of the first permanent molar.
At two years: the incisors show signs of wear; the second permanent molar will appear (now present are three temporary and two permanent).
At two years six months: the central temporary incisors are cast, and the first pair of permanent incisors begin to protrude through the gum.
At three years: the two permanent incisors are through, and two further permanent molars appear, pushing the temporary teeth in their position out of the way. It is only now that you can see whether your youngster has a correct mouth, that is, the upper and lower teeth exactly in line and no overshot jaw (parrot mouth, see p109).
The teeth continue to change throughout the horse's life, and it is not until he is between **four and five years old** that the **last two permanent molars** are cut.

The development of the teeth

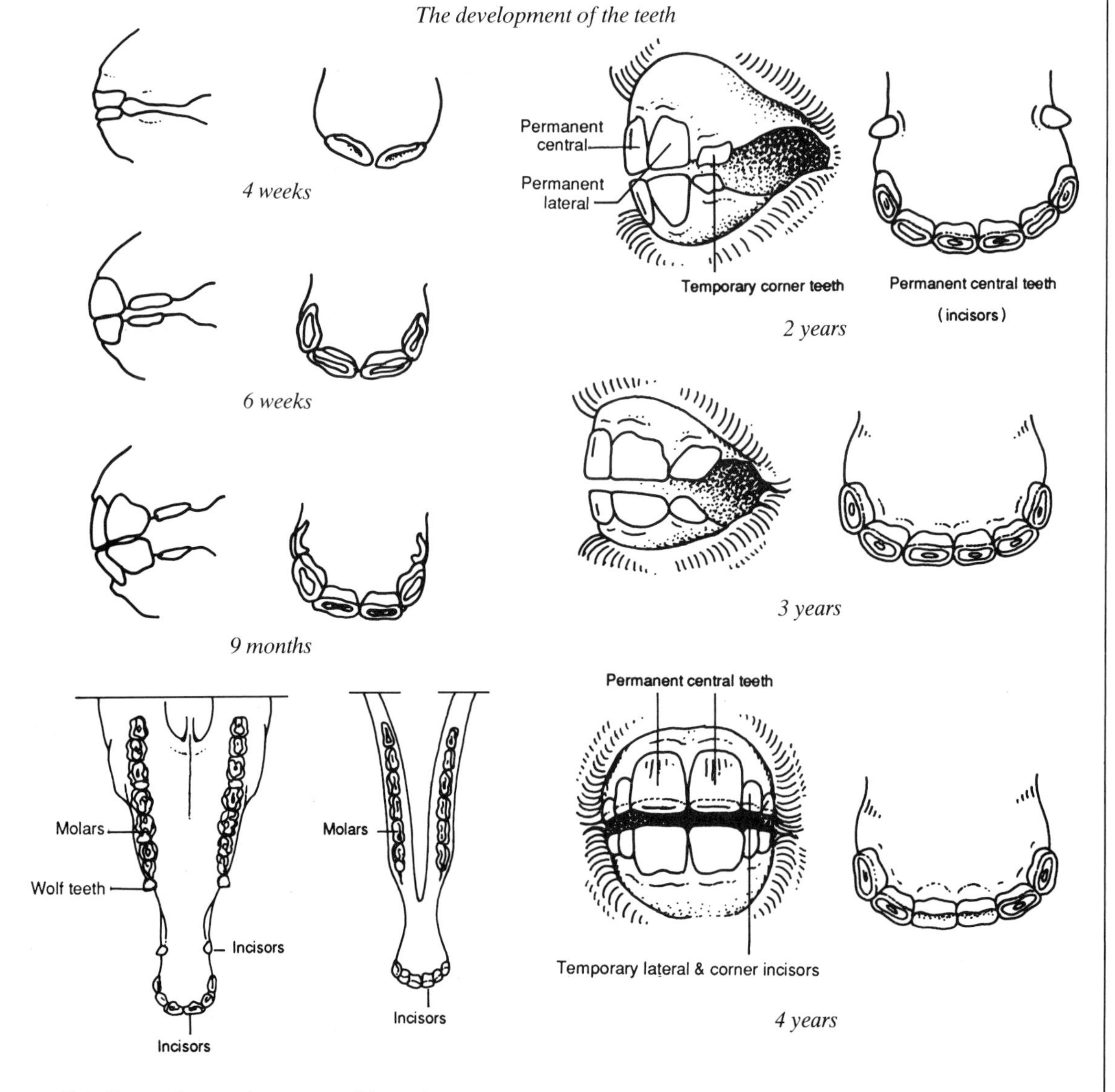

Grinding surfaces – the upper and lower jaws

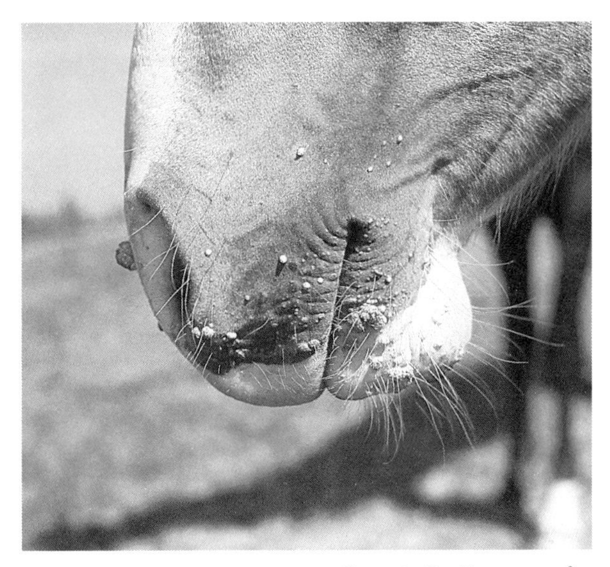

Warts seen on a yearling; these will magically disappear after a period of time

sort known as milk warts and there is no cure for them, but they will eventually disappear, either when there is a frost or during the hotter months of summer. The old stud grooms always maintained that warts in the youngster were due to a deficiency of lime in the soil; however, there is no scientific evidence to prove this and they are more likely the result of a virus passed from one youngster to another as they graze.

Milk warts look very ugly, and if the youngster rubs his face they will bleed. Be patient, I have seen the prettiest of pink noses magically smooth again with no resultant scars, even having shed a number of large, ugly warts. If you are concerned about warts, consult your vet.

The second winter

Once autumn sets in, the youngstock of most breeds need to be in at night for the duration of the cold months, to ensure optimum growth. Bring them in when the nights become very cold or wet; alternatively if the weather remains kind, you may well decide to leave them out at grass for a while longer, though they may need some additional feed in the form of good hay.

During a young horse's second winter, regular feeding **together with the appropriate vitamin supplements are as important as during his first winter** – his growth activity is still great (see chart 2, p140). Rugging up a youngster during winter, either at night, or when he is out at grass during the day, is not necessary. It is far better to allow nature to take its course and he will grow a good thick winter coat. At this age he will still be playful – rugs can easily get damaged, and can be a danger. All youngsters thrive on routine, so make your plan and stick to it. Remember that regular worming, visits from the farrier and annual vaccination are of paramount importance to ensure optimum health in the young horse.

SUMMARY: THE YOUNGSTER'S SECOND WINTER

1. Regular twice daily feeding incorporating a suitable vitamin supplement.
2. Observe normal behavioural patterns: if you are familiar with these you will be more likely to notice if anything is wrong.
3. *Worming:*
   Yearlings: every 4 to 6 weeks
   Two-year-olds: every 6 to 8 weeks
   Alternate the type of wormer. Ensure that at least once a year you use a wormer which is effective against bot worms *and* larvae, also one that expels **tapeworms**.
4. *The Farrier:*
   Yearlings: trim every four to six weeks
   Two-year-olds: trim every six to eight weeks
   If feet are showing signs of splitting, seek advice from your blacksmith; front shoes may be necessary.
5. Annual vaccination.

Regular worming is of paramount importance for the well-being of youngstock

SHOWING THE YOUNGSTER IN HAND

In-hand showing is a great education for the youngster but it is not a necessity. Yearlings in particular can be likened to a tall, awkward teenager, and there is no point in exhibiting any young horse if he does not look good. However, you may find that towards midsummer when he has had the best of the spring grass he looks better; perhaps you want to take the opportunity of showing him as part of his education – it is also an ideal way of seeing how he compares with others. (See also Showing the Mare and Foal, p115.)

The art of showing in hand is something that no manual can teach; it comes from years of experience and knowledge of how to prepare a horse for the show ring. If you are an amateur, but have a top quality youngster and want to see him win at County Show level, it may be best to seek help from someone who is an experienced exhibitor. Good feeding and sensible handling for some weeks beforehand, besides actual presentation, are all part of the secret of winning in hand. Nevertheless, the art of showing in hand can be learnt by most people – and remember you do not have to be a competent rider to start off with. Observe how the best competitors have their horses standing four-square, and how they run them up; then work out the best way to gain an understanding with your own and do it for yourself.

Feeding, Strapping and Condition: A youngster that is to be shown in the ring may have to be fed extra rations – the exception to this rule being most of the native breeds – but *do not* allow him to put on excess fat; it is not good for the young horse to carry extra weight as his bones are still forming. If you decide to give extra feed do so carefully according to the time of year and the build of your youngster. Think ahead and observe his condition two months before the show – that is the time to make your plan of action; and remember, regular strapping is always beneficial. His condition should be naturally good, and no amount of last-minute use of expensive shampoos and conditioners will help. If your youngster looks well, there is no reason why you can't bring him in from grass the day before the show.

What the handler should wear is discussed in Chapter 6, p116.

Showing the Yearling

Type, breed and temperament will determine whether your yearling should be bitted for in-hand showing. Most ponies, except very boisterous colts, can usually be shown in a leather headcollar or filly slip with a light leather or tubular web lead-rein; however, there are not many yearling colts of Thoroughbred or similar type who are sufficiently well mannered not to need the restraint of a bit, and these should be shown in an in-hand bridle. **Except for the native pony types**, most breeds should have their manes plaited, and the tail may be plaited too, but this is a matter of preference; fetlock feathers should be trimmed. Manes will almost certainly need pulling; most youngsters dislike this operation, primarily because it hurts them. Nevertheless, at some time before it gets to the stage of being too long and becoming knotted, it should be pulled, so find an assistant and work at it gently. If you have a major problem I suggest using a twitch, which does have a tranquillising effect and will allow you to smarten up the mane quickly and efficiently. All in all, your youngster should look smart and clean, with a glossy coat and hooves suitably oiled. Vaseline around the eyes and nose will help with the overall 'chic' appearance that is expected in the show ring.

In order to show the yearling with any degree of success, he must lead obediently in hand, he must understand voice commands, and should generally respect you. If he has not already been taught to tie up, then he must be taught now: attach a piece of string to the tie-ring, and tie the lead-rope to the string in case he pulls back – if he continually pulls back and evidently finds breaking the string quite fun, you will have to resort to a racking chain. *Do not* attach the chain to a piece of string, which is a dangerous practice – if a horse pulls back so the string breaks, the chain will be flung into his face. A chain is used as an alternative to a rope to prevent him from pulling back, but *be careful* when you first use this method. You will need to be at his side constantly to ensure that when he finds he cannot break the chain

Never lead a horse with a chain – if he pulls away your hand could be badly damaged

by pulling back you are there to assist if he panics. Never lead a horse from a chain as your hand could be badly damaged if he pulls away. **The main objectives of the leading stage are as follows:**

1 To teach the youngster to respond to the voice commands;
2 To teach him to go forwards, and to stop when requested to do so;
3 To show him the outside world by introducing him to objects he would not normally see.

Remember the horse is essentially a herd animal so he can become very anxious when separated from his friends, especially when he is still very young.

Bitting: This is the time to bit your yearling if it is necessary – unless he is a pony type, when he will not need a bit until he is older. Yearlings have very sensitive mouths; their skin is still young and the bars of the mouth very soft. Choose a straight-bar, loose-ring bit, preferably nylon, rubber or vulcanite, with not too thick a mouthpiece. First, put the bit on in the stable and leave him for an hour to play with it. It is not necessary to tie him up. Do this daily for a few days, and lead him from the headcollar with the bit still in his mouth. As he becomes accustomed to it, begin to lead him gently, from the bit using a brass or leather coupling which attaches on either side of the bit and will therefore give an even pressure, even though you lead from one side. Use your common sense: before you go out to a show, ensure that he is familiar with the bit, and that you have allowed plenty of opportunity at home for him to accept this new object in his mouth.

It is not necessary to bit a yearling, unless you are going to show him, *or* he has become boisterous or stroppy at home and difficult to lead. I suggest you use a Chifney (also known as an anti-rearing bit) attached to a slip-head to give you better control (although note that this bit is not acceptable in the show-ring).

Practising at home: 'Practice makes perfect' – but do not overdo things and sicken your youngster. As a rule, about six weeks to a month before the first show start bringing him in every other day and practice leading in hand. Use the voice for commands, and always carry a stick. As much as the walk, practise standing him up square so that all four legs are equal, and teach him to remain so while you stand back; in the ring a judge can then view him without you in the picture. Praise him when he does well. Ensure that he remains happy and settled with people all around him; when a judge approaches him he is then more likely to stand still without fidgeting. Keep an eye on his feet during this preparation period – if you are walking him on roads he may get footsore and need shoeing in front.

Once the walk and stand are established, the trot-up has to be mastered – the trot is of paramount importance, as this is the only time that the horse's action is demonstrated individually in the ring. To show off the pace to its best, **the handler must learn to run in time with the**

(Opposite above) *Introduce a yearling to poles; this adds some interest to the in-hand work*

(Opposite below) *Incorrect trot-up in hand; the handler is too far in front of the youngster's shoulder*

(Right) *Correct trot-up in hand*

youngster's stride; lightness and rhythm, and also *extension* is required, especially when showing a warmblood. The horse must run freely with the handler positioned just behind his elbow, and to achieve this successfully the handler must move well enough himself to show off his youngster's paces to best advantage. **It is important, too, that the head and neck are straight** and not pulled round towards the handler. If the horse is constantly led with his head turned in, the muscles on the lower side of his neck will develop more fully than those on top, which is incorrect and difficult to rectify. Some professional showing people avoid this situation by long-reining instead of leading in the traditional way (see p163).

Loading: It is important to practise loading before the show. There is nothing worse than being well prepared for a show outing and at the critical time of departure your youngster refuses to load; a fight at this age is not to be recommended! Check the condition of the lorry or trailer carefully, as youngsters have a tendency always to find that sharp edge or protruding nail; also make sure that the vehicle is clean and disinfected as viruses can be picked up easily. I suggest you park the lorry or trailer on grass to practise loading. If using a trailer, open up the central partition to allow more space, also the front unload if it has one, to allow in more light. Take your time, and have a bucket of feed ready to coax the horse up on to the ramp. Always have at least one other person available to assist, as you may need to use the lunge-line.

Once your youngster is in the trailer, feed him so that he feels comfortable. If it has taken a while for him to go in, practise more than once *except if* the lorry has a very steep ramp and unloading is difficult. In this case it may be best *not* to over-practise – use your common sense, and assess both the situation and the type of temperament your youngster shows.

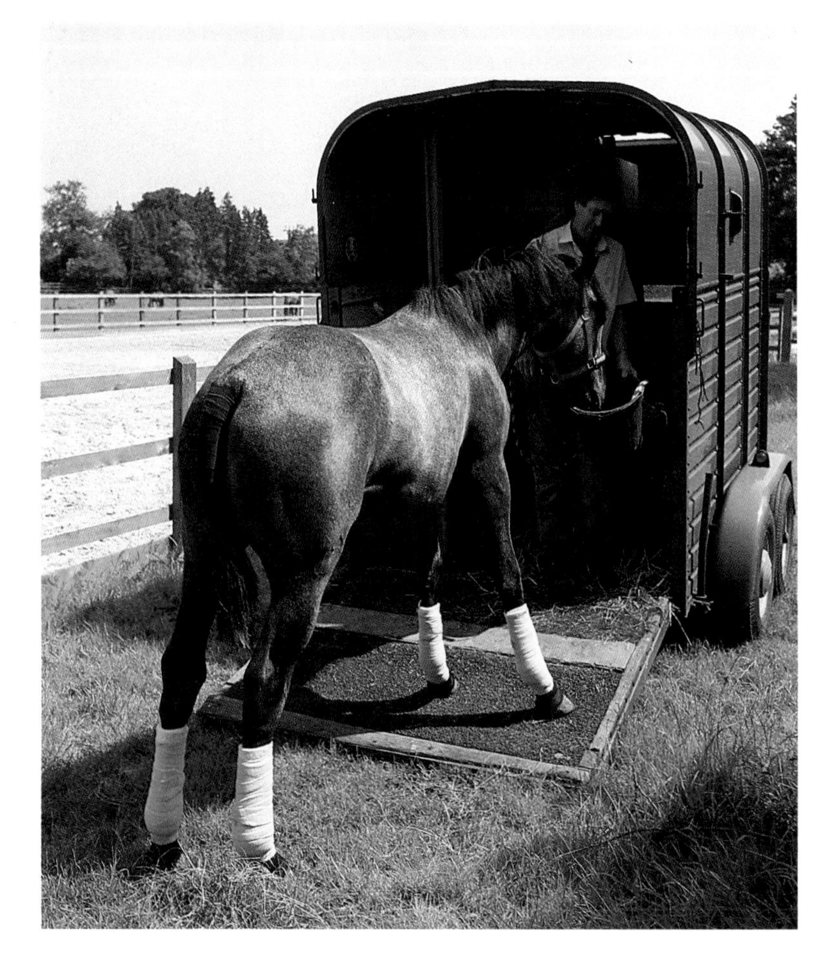

(Left) *To practise loading a young horse, open the trailer's centre partition as wide as possible, or take it out altogether; if it is front unload, open the front top door too. Offer the youngster some feed, at first on the edge of the ramp, then gradually encourage him to bring his forefeet on to the ramp and so slowly into the trailer. Take your time: do not rush him*

(Below) *If he resists and it is obvious that he does not want to walk in, use a tubular web lunge-line (not nylon as it will burn both your hands and the horse's skin) and fix it to one side of the trailer*

(Opposite above) *One person should remain at his head still encouraging him to walk in with the bucket of feed, whilst the second person takes a firmer hold on the lunge-line behind his quarters. This person must be experienced and should know exactly when to slacken the rein if the horse resists and starts to go backwards*

(Opposite below) *Once the youngster is loaded, do not be in a hurry to close up the ramp. He should be encouraged to stand inside and finish his feed before he is unloaded*

Showing the two-year-old

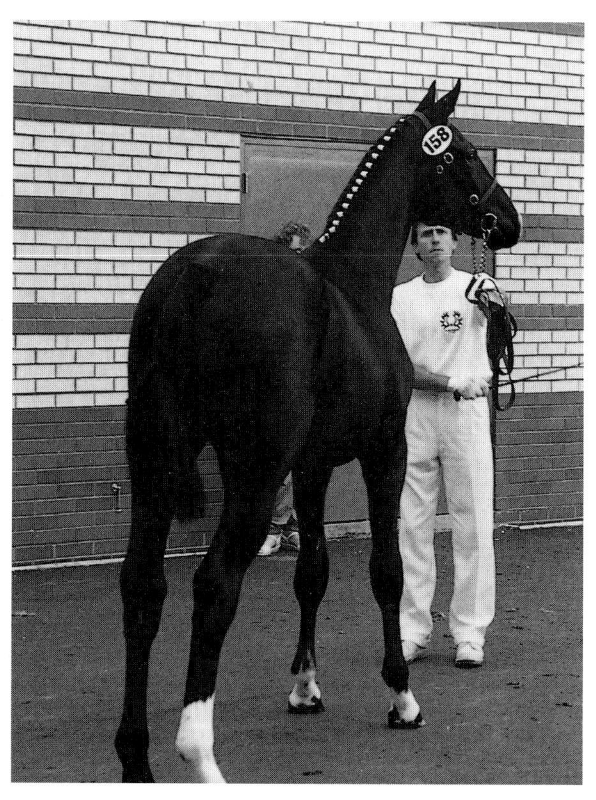

By the time your youngster reaches the summer of his second year he will have filled out, and both his physique and mental attitude should have matured; it is therefore more likely to be a better year to show him off in the in-hand ring, and no doubt he will remember whatever lessons you taught him the previous year.

Preparation is more or less the same as for a yearling (see p150) except that because he is older and stronger you can do more. He can be introduced to a roller, and eventually side-reins. When first putting on a roller use your common sense and do the girth up gently, then walk and trot him with just the roller before attaching the side-reins to the bit. If you do decide to use side-reins at this stage be sure to adjust them quite loosely and do not force him into an outline. *Never* attach the side-reins to the bit except in a manege or small field, so that if he does not like the pressure and plays up, you are in a relatively

(Above) *A suitable turnout for in-hand hunter classes*

(Right) *Long-reining is an excellent way to prepare for the show ring*

(Opposite above) *A yearling's mane in need of attention*

(Left) *The correct turnout for a warmblood show*

confined area and it is therefore easier to persuade him to go forward with lunge whip at hand.

I do *not* recommend a lot of lungeing at this stage unless your youngster is of a large breed or extremely well developed, however it is an important lesson in discipline which in addition teaches the youngster to listen to your voice (see p162). Even if your youngster is well-mannered and docile at home, he is bound to find his first outing very exciting; so put him through his paces when there are other horses around so that he becomes accustomed to hustle and bustle whilst in a disciplined situation himself. If you take him out and he behaves badly, **blame yourself, not the horse**. Go home and do your homework – maybe your two-year-old needs a more interesting and varied work programme.

At the showground

Make sure that you arrive in plenty of time, and always take an extra pair of hands: you should not leave a youngster alone in a lorry, and you will need someone to help prepare him before entering the ring. Try and choose a quiet in-hand show for his first outing, rather than a large county show with wild activity in the form of mad ponies, show jumpers and funfairs! I wish you lots of luck. Provided that your preparation has been done consistently you should have no problems; but if you do, take your time – and strange though it may seem, some youngsters settle down once they are in the ring with others. Only chastise your youngster if he genuinely steps out of line.

SUMMARY: SHOWING YEARLINGS AND TWO-YEAR-OLDS

1. In-hand showing is not a necessity, but it is undoubtedly an education in itself for your youngster.
2. Decide if he needs extra feed at least six to eight weeks prior to the show, and establish a routine.
3. Practise his 'walk', 'trot' and 'stand' some weeks before the show.
4. Practise loading.
5. If he needs a bit, ensure that he has happily accepted it during practice.
6. Make sure that he is content with people around him, and that if the judge asks him to pick up a foot, he will do so willingly.
7. Make sure that he looks well and is in good condition; **otherwise do not show him.**
8. Ensure that you have the correct bridlework and that your own dress and equipment are ready.
9. Read the show rules and regulations, and make sure that his vaccination papers, if required, are up to date.

THE THREE-YEAR-OLD

In his third year your youngster will show significant signs of the physical and mental development more typical of the adult horse.

At last his physique gives a better idea as to how he is going to turn out as a mature horse: growth development has slowed down.

He will undoubtedly show a more mature attitude towards life, and by the end of spring will be ready to continue with further training.

This chapter discusses various methods of training from the ground, also the initial education required for those youngsters with a future as potential performance horses, in whatever sphere; there are different schools of thought on the subject, so the basic requirements for a varied preparation are considered.

As always, it is important to understand and observe the young horse, and to continue a well balanced feeding programme together with a regular worming programme, and visits from the farrier.

As a three-year-old, the young horse is by no means yet 'full grown' but he is on the threshold of adulthood. His further training is beyond the intended scope of this book so we will leave him at this undoubtedly exciting stage, happily backed and with a full and challenging future ahead.

EARLY MANAGEMENT

Intelligent observation, good psychology and quiet discipline are the responsibility of the young horse's trainer. If you are at all worried about handling your youngster, particularly if he is stroppy or pushy to the extent that you cannot manage him, then it is better to admit it and get assistance. There is no need to be embarrassed: all youngsters go through difficult stages, and if they did not there would be something wrong; moreover not everyone can handle them at these times. Remember, too, that smell is a powerful factor in the horse's communication process, and a horse can smell fear, or anxiety on the part of his handler.

The un-touched three-year-old

A young horse that has been left to 'grow up' without being handled needs to be 'tamed' in the same way as would a youngster straight off the moors. The end result should be no different from the horse that has been handled since birth, but the process of 'taming' is time-consuming and tough for both the horse and trainer, and there may well be more fights and more corrective treatment involved. The older the horse is, the more difficult the task, mainly because of the animal's size and strength and also because it is accustomed to having its own way. However, the horse has been domesticated for the last 6,000 years and is obviously quite capable of accepting Man as the dominant being. It is mostly a social animal, and its trust comes from the building of a relationship; it is therefore important to establish a bond with your young horse. The various stages of taming are basically the same as those involved when handling any youngster and thus similar to those already discussed. To summarise, a 'taming programme' might be carried out in the following order:

1 Confine the horse to a stable.
2 Accustom him to human presence and teach him to associate feed as a reward.
3 Teach him to accept touch, so that eventually he will allow you to groom him.
4 He will need to be haltered and taught to tie up (see p150).
5 Teach him to pick up his feet, so that ultimately the farrier, too, can attend to them.
6 He needs to be led out on a headcollar, and to be taught the voice commands with the aid of the whip to encourage him forwards, and with feed as a reward.

Once you have achieved this, your un-touched youngster will have been tamed. However, be under no illusion: it will take a great deal of time, patience and understanding on the part of a competent horse person.

EARLY TRAINING

When you back your youngster and how you go about it will depend on your circumstances and your horse's physique; there is no immediate hurry, but on the other hand do not put it off for too long. Mentally, most youngsters of three years old are ready to do some work and will benefit from a structured period of disciplined learning. No great harm will be done if a horse is not backed until his fourth or fifth year provided he has been well handled in hand, but the breaking process could be more difficult as he will undoubtedly be stronger and probably less willing to co-operate when introduced to a totally new discipline. If you have a very big, strong young horse it really is advisable to back him in his third year. Whatever you do, when he reaches this stage you will have to decide the following:

First, are you going to prepare your youngster for breaking, and back him yourself? If so, do you have the required facilities and assistance?

Second, are you going to prepare your youngster for breaking, but send him away when he appears to be ready to back? In this case the length of time required by a trainer will be approximately four weeks (if you have done a good job).

Third, are you going to send him away to a trainer for the entire preparation and ridden activity? The time required by a trainer for this will be approximately six to eight weeks, though it will be dictated by the youngster's ability to learn, and his co-operation during this period.

A well fitting lungeing cavesson using the nose-piece as a drop noseband

A well fitting lungeing cavesson

If you have made up your mind to send your youngster to a trainer, do your research first. Find a competent person whose methods and ability you believe will suit your horse's temperament, and someone whose horse management, knowledge, understanding and patience you can trust. He should be a good observer, and when he encounters difficulties he should know why he has done so, and have an effective solution. You should be happy with the yard, its facilities and the proposed fees.

If you decide to keep your youngster at home and break him yourself, it is important to have the necessary equipment and help. You will need at least the following:

1 An area where you can lunge, preferably a small paddock or manege
2 Lungeing cavesson; the Wels lungeing cavesson as used at the Spanish Riding School in Vienna is to be recommended
3 Two tubular web lunge reins
4 A bridle with a kind snaffle bit and flash or drop noseband
5 Side-reins, leather (or nylon) without elastic inserts (as these encourage the horse to lean on the bit)
6 A lungeing roller with two Ds for the attachment of side-reins (this is not, however, a necessity, as you could use a saddle)
7 Set of brushing boots
8 An assistant who is light in weight and sufficiently competent to back a young horse.

A three-year-old should be out at grass day and night in the early part of the year so that he may benefit from the spring grass. Once the goodness has gone out of the pasture he can be brought up at whatever time is convenient for you or your chosen trainer, as long as at least six weeks are allowed for preparatory groundwork and for backing. Once you have made a start, it is preferable to stable your youngster for a good part of the day; this will give him a set routine which will support the work that is being carried out. The stable will eventually become his home where you hope he will feel relaxed and comfortable, so ensure that he has company nearby and is not isolated. Otherwise vices caused by frustration such as weaving or box-walking that have so far been avoided could begin now. In my opinion **he should not be stabled 24 hours a day;** young horses *need freedom* and he will be much more amenable if he has *at least* **eight hours out at grass every day.**

If you are in the least doubtful as regards your ability and/or facilities to break your youngster, send him to a professional rather than risk subjecting him to a bad start.

Preparatory groundwork and backing

The following six-week programme describes a routine which prepares the young horse for backing; it is based on the assumptions that a) the horse has been well handled as a youngster, and that as a three-year-old he knows how to tie up, to stand and to walk and trot in hand, and b) that the owner or trainer knows how to lunge and school horses. It is important that throughout this relatively concentrated training period you tie him up, groom or brush him off, and pick out his feet *every day*. Spend time with him, get to know him and his characteristics: this is the time when the foundations of a real partnership are laid.

Arrange for the farrier to shoe him in front so that he can be walked out on the roads.

The following exercises should be carried out for 20 minutes to 3/4 hour, roughly 5 days a week, depending on the horse's response.

WEEK ONE

- Introduction to bridle and bit (see p151)
- Basic groundwork: walk, halt and trot in hand
- Introduction to a roller (or saddle if none available) and side-reins (never attach side-reins to the bit until the horse is in the area where he will commence work)
- First lessons in lungeing (assistant will be required) – maximum 15 minutes.

Remember to praise the youngster when he has done well – a Polo mint or cube is usually found quite acceptable!

The main objectives of lungeing

1 To teach your youngster the voice commands, obedience and further trust

2 To ensure that he respects your wish for him to go forward at the walk, trot and canter, and to make smooth transitions from one pace to another

3 To assist him in maintaining rhythm, balance and tempo

4 To help develop his physique and muscle structure

5 Eventually to engage the hocks and increase the driving power of the hindquarters

Always wear gloves when lungeing and **carry a lunge whip.**

Lungeing, like in-hand showing, is an art and should be an integral part of the horse's education. A youngster should be good to lunge prior to backing, and if you are not experienced in lungeing technique then it is advisable to seek help.

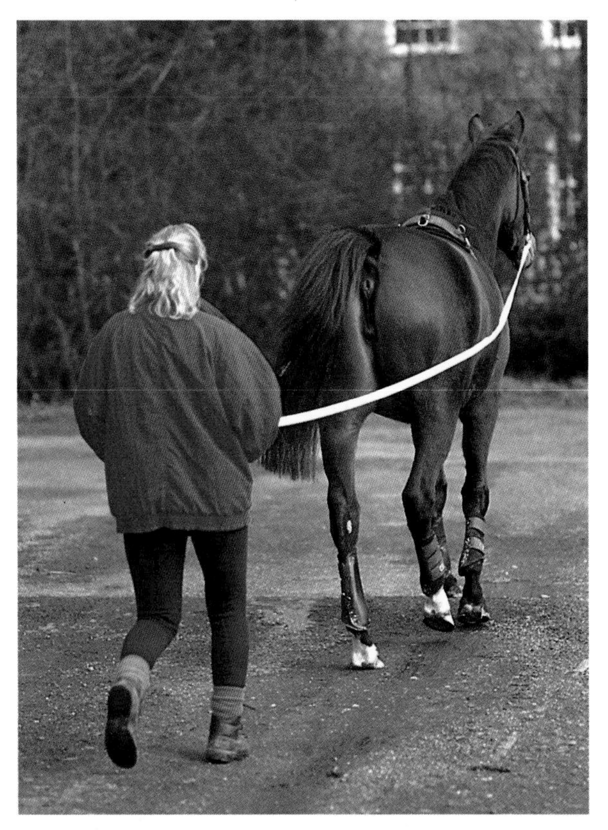

Long-reining, a view from behind. An exercise which encourages the young horse to go forward on his own and take contact on the bit whilst going in a straight line

Lungeing with a saddle, the stirrups left down to accustom the young horse to objects flapping on his side

WEEK TWO
- Continue lungeing lessons
- If roller used, now introduce to saddle
- Gradually introduce to cars by walking in hand and observing traffic from a safe distance
- Lead over poles
- Lunge with saddle (remove the stirrups or ensure that they are well secured and will not slip down)
- First lessons in long-reining (assistant will be required)

The main objectives of long-reining
Long-reining is not always considered an essential exercise prior to backing a young horse; however, the practice of long-reining in addition to lungeing serves a very useful purpose in his preparation:
1 He learns to take contact on the bit, and to go to the right or to the left from a rein-aid
2 He learns to take a contact on the bit without a side-rein or a person by his head
4 He learns to go forward on his own
5 Long-reining gives more variety to his work, and acts as an additional exercise of obedience and education

Long-reining is a skill that needs to be learned. In effect, the horse is driven by the trainer on the ground – he used two reins which pass through the lower Ds on the roller; **no side-reins are required**. The trainer must be careful to be light in the hand and not allow the youngster to overbend. A lunge whip is held and the horse tapped lightly from behind if he does not go forwards in response to the voice command.

To begin with, an assistant should attach a short rein to the headcollar and walk alongside the horse's head; as the command is given from behind, the assistant walks forward. Once the horse gets the idea, the assistant can detach his rein and eventually disappear out of sight.

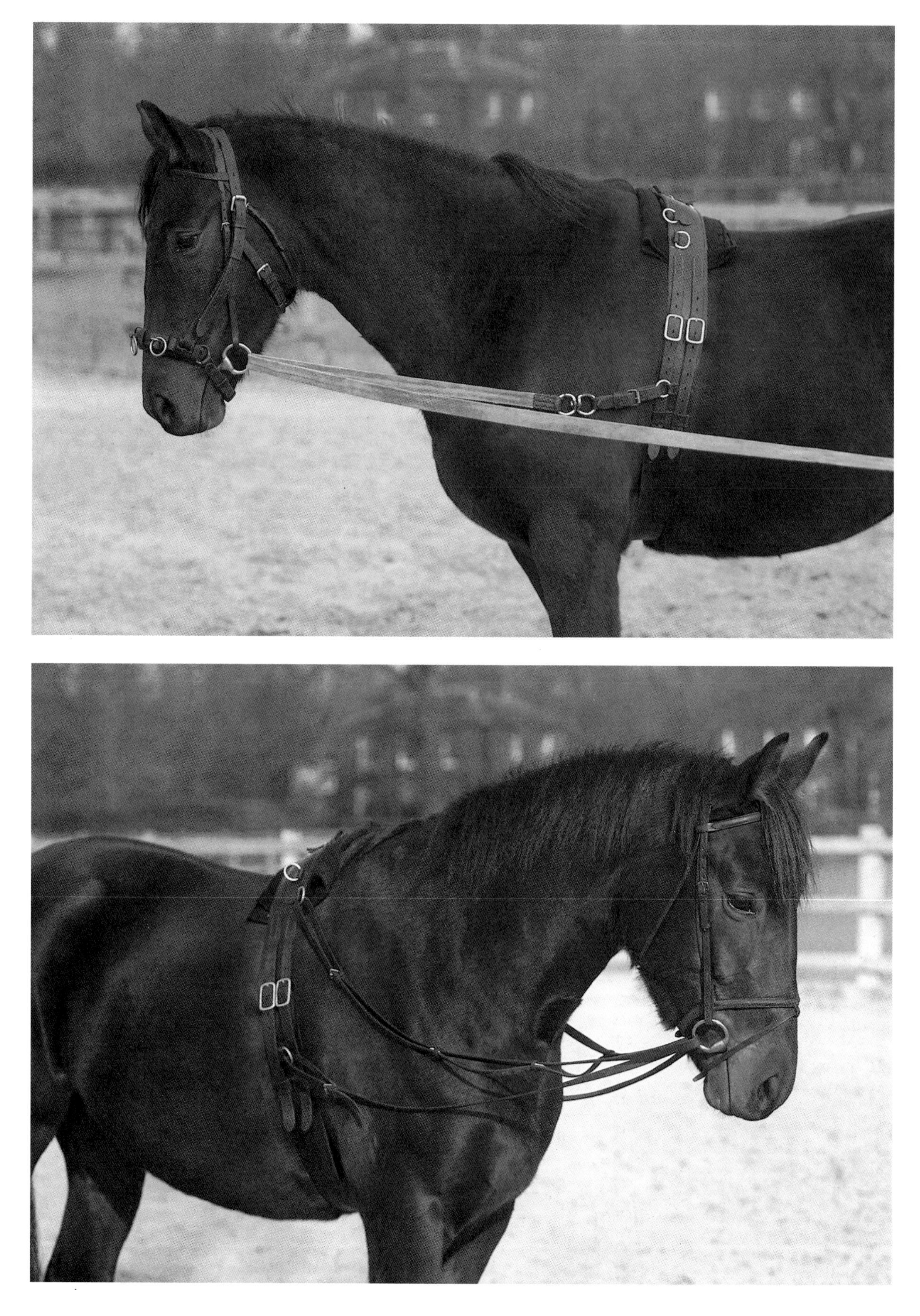

Alternative ways of using equipment for lungeing purposes if problems arise. (Opposite above) *The lunge rein is attached to the side of the roller and passed through the bit. This will restrain a very strong youngster which attempts to charge off away from the circle. However, it can make an undesirably strong contact on the inside rein*

(Opposite below) *Vienna reins. One rein passes from the side 'D' on the roller, through the bit and up to the 'D' just below the wither. By using these reins the horse cannot lean on the bit*

(Above) *Vienna reins used in a different manner: each rein passes from the girth, between the forelegs and up to the bit. Using the reins in this way will make the horse work more from behind, but they should not be used on young horses with an immature muscle structure*

(Left) *The lunge rein attached to one side of the bit, passed up over the poll and then down through the other side. This is not recommended for young horses as it is not necessarily giving an even pressure on either side of the bit*

WEEK THREE

- Vary the above; do not sicken the youngster
- If lungeing well at the walk and trot, pull down stirrups on the saddle
- Improve his reactions to your voice
- Carry out his work with another horse or horses working at the same time, in the vicinity
- Lunge over poles
- Long-rein away from the yard; use fields or lanes
- Teach him to loose-shool

Loose-schooling:
To teach a horse to loose-school requires at least two people on the ground and is easier to accomplish if the young horse knows what is expected of him on the lunge, and is familiar with the voice commands. Loose-schooling has the great advantage in that there is no interference to the youngster's head, as the trainer has no contact with him.

Begin by using a relatively small area (approximately 30 metres square) – if you let a youngster loose in too large an arena you cannot expect to have sufficient control and he will not understand what is required of him. Trainer and assistant should each have a lunge whip, and to start with the horse should be first trotted around in hand, and only then let loose and left to continue the circuit alone, encouraged by the other person with the help of his whip. If the horse stops and turns round, do not let him continue – he must learn that he is expected to keep going in the same direction until asked to change. Once a horse becomes familiar with what is required of him, you will find that this is an exercise that he, and horses of all ages thoroughly enjoy, and it will be an achievement when he reacts to your

In the first stages of backing, the rider is given a leg-up so that he may lean across the horse's back so letting the horse become accustomed to the weight

It is not advisable to put weight in the stirrup at this stage. Note that the rider should always wear a back protector as well as the correct headgear when breaking in a horse.

voice alone. The intelligent youngster will also learn to change the rein on request.

Do not forget to reward your horse when he has done well, and remember that loose-schooling is teamwork, and that includes your assistant – if the latter is merely a friend who has called by to give you a helping hand, make sure he knows what you require of him, otherwise you can easily confuse the young horse.

WEEK FOUR

If all is going well and you have overcome any problems, he should be ready to back. If not, continue as in Week Three; also it may be advisable to do two short periods per day.

Backing the young horse:

Lunge first, then if he is settled ask your assistant to lean across his back, first at a standstill and thereafter in walk; approximately five to ten minutes will be sufficient for the first couple of times. By the end of the week the rider should be able to sit on him in walk whilst still on the lunge.

Do not bore your youngster. Vary your lessons, and as a result he will be more co-operative.

Do not have a fight, unless you are sure that you are going to win; it is far better to use good psychology and work around the problem, or seek help.

WEEK FIVE

- Short lunge period and then continue with rider in walk and trot
- Rider loose in arena, off the lunge
- Ride away from the premises with someone on the ground at his head to give him confidence
- Loose-school over small jumps
- Continue with all previous exercises

(Above) *If the horse seems quite happy with the extra weight, he should be led a few steps with the rider's body leaned across his back*

(Left) *The rider is quietly in position as the youngster is led a few paces*

(Opposite above) *Note that the rider's hands remain quite still as the young horse shows some resistance to going forwards. A neckstrap should always be used on a young horse at this stage of his training*

(Opposite below) *A few days later: the youngster is going nicely forward on the bit; he should now be ready to go loose without the lunge-line. At first the person on the ground should remain in the same position and continue to use the now-familiar voice commands, thus continuing to give the youngster confidence*

(Above) *Two young horses being worked side by side on long-reins*

(Opposite above) *A young horse in harness being lunged, to accustom him to blinkers in preparation for driving*

(Opposite below) *A young horse being taught how to pull, ready for the vehicle*

WEEK SIX

- Continue all lessons, varying the exercises
- Youngster may be ready to hack out for a short period with a lead horse

Do not overdo the following exercises with a three-year-old:

1 Lungeing on a circle – a youngster's balance is not yet secure, and his muscle structure must be built up gradually. Too much lungeing tends to make work in straight lines more difficult.
2 Do not jump the three-year-old and rider over anything above 2 feet (60cm): his bone structure is not yet sufficiently strong. Loose-jumping, however, is a good exercise.
3 Do not lunge or work a youngster on hard ground: a splint may form as a result (see p146).

Summary

The above programme is merely a guide and each youngster will differ according to temperament, intelligence, type and ability. Common sense and experience on the part of the trainer will determine an appropriate routine. As a three-year-old, once your young horse is backed and happy to walk, trot and canter with a rider, and perhaps gently hack out with a companion, that is really all that need be accomplished at this stage. However, you will have to decide whether you interrupt his training now, or continue to ride him two or three times a week to keep him going until the winter.

It may be advisable to 'end on a good note' and do nothing more until the spring of his fourth year. However, this decision will depend on your personal circumstances and the horse's disposition – for example if the youngster is for sale, then it may be better to keep him going so that a potential customer may see him ridden and try him out. If your three-year-old is highly intelligent

and becomes easily bored then he will probably benefit from continued light work. *Horses do not forget.*

It is early days to decide the future of your youngster, but by this time you should have an idea of what he is likely to be good at, regarding performance. His pedigree is, of course, highly significant – but even if he has been bred, say, for dressage, that doesn't mean that he should only go round an arena in circles. All horses and ponies can jump; some make a better shape over an obstacle than others, and some may be brave and others not so brave; but every youngster should be given the chance to learn. Perhaps he has eventing bloodlines, in which case he should certainly be introduced to all three disciplines required in horse trials. But even for the dressage horse, variety is all-important, and hacking out the youngster enables him to acclimatise to the much wider world. Always hack out with another horse, and preferably one which is calm and obedient and good in traffic – there is nothing worse than a lead horse that refuses to walk once in the woodland, and when asked to canter, disappears at a spanking gallop. Avoid this sort, and try to find a 'nanny' who will be a good example to your own willing youngster. Bad habits are easy to learn and difficult to forgo.

In conclusion, perhaps the most important thing to remember is that **any young horse needs variety** – otherwise he will quickly become sour and that is when problems arise. It is your duty as the owner of a youngster to introduce him to as many aspects of equestrianism as may suit him. Try not to be blinkered in your outlook; give your youngster interesting work combined with enjoyment, and do not be afraid to work him amongst other horses – he will like the company. However, above all, remember that a three-year-old still has a lot more growing to do.

STAGES OF GROWTH FROM FOAL TO TWO YEARS OLD

BORN: April SEX: Colt BREED: Hanoverian

3 weeks: *A colt foal (AB) seen here with his mother in May, a time when good spring grass ensures plenty of milk, resulting in maximum growth potential for the foal*

12 weeks: *Already a fine strong colt which shows presence as well as good movement*

Use a well fitted leather headcollar if you travel the weanling

(Opposite above) 5 months: *He is now quite old enough to wander away from his mother for considerable periods of time and make good pals with other foals of a similar age*

(Opposite below) 6 months: *Weaned and ready to move home if circumstances change. When re-locating a weanling use the following priority checklist:*

1 Ensure vaccinations are up-to-date

2 Note date and type of last wormer administered

3 Take note of the current routine and feeding programme, and endeavour to keep to it – any changes of feed should be gradual

(Left) *AB meeting one of his new companions prior to being put out into the field. When putting a weanling out with a group of foals which already know each other well, it is recommended to introduce the new foal to one of the group beforehand. In this way he may be accepted more readily*

(Below) 8 months: *AB observed grazing contentedly with other youngsters of the same age-group*

(Opposite above) 10 months: *Having fun out at grass after a night in the barn, showing how important it is for youngsters to be with others of a similar age-group*

(Opposite below) 11 months: *It is not necessary to rug a yearling during winter, but it is worth introducing him to a rug. Note that a yearling will not carry much weight at this stage of his growth, though height increase is substantial (see Growth Chart p140)*

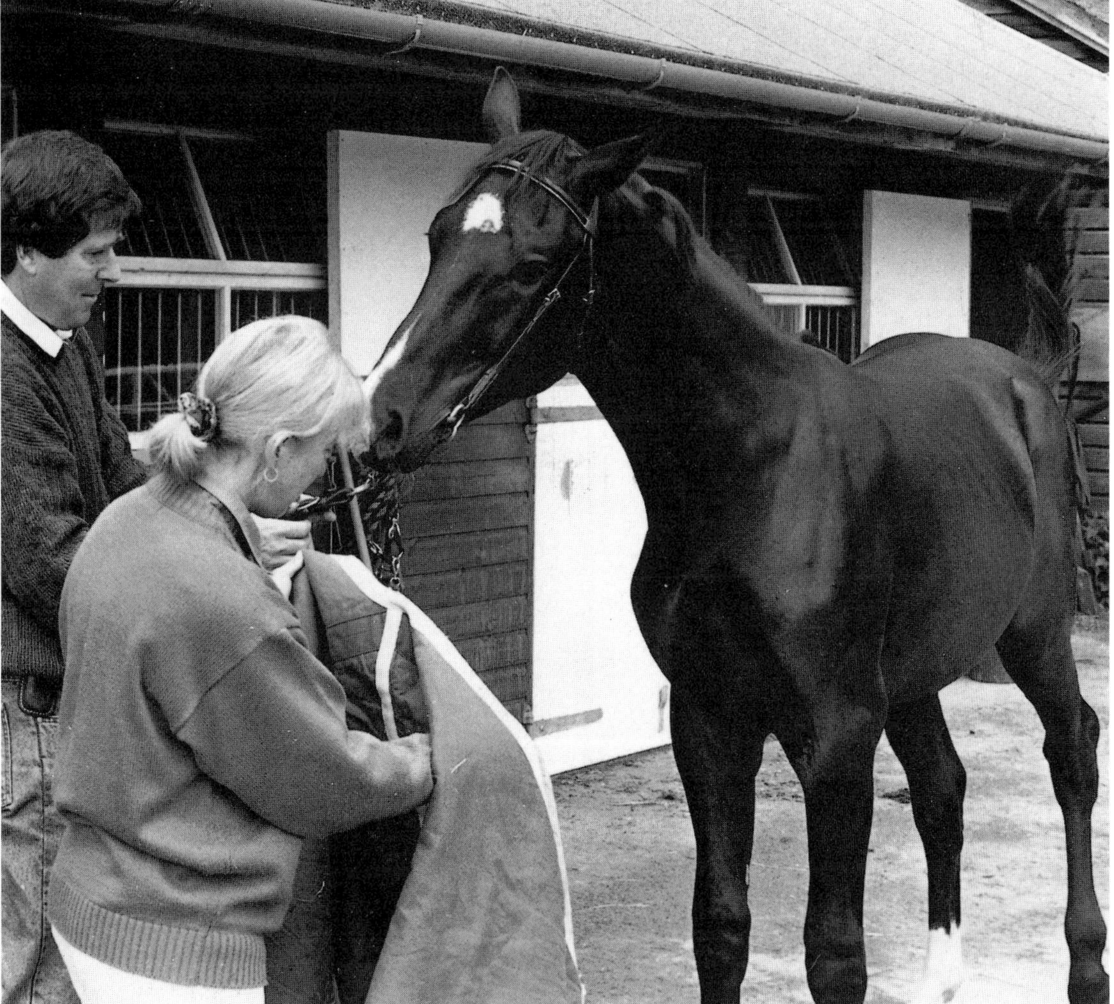

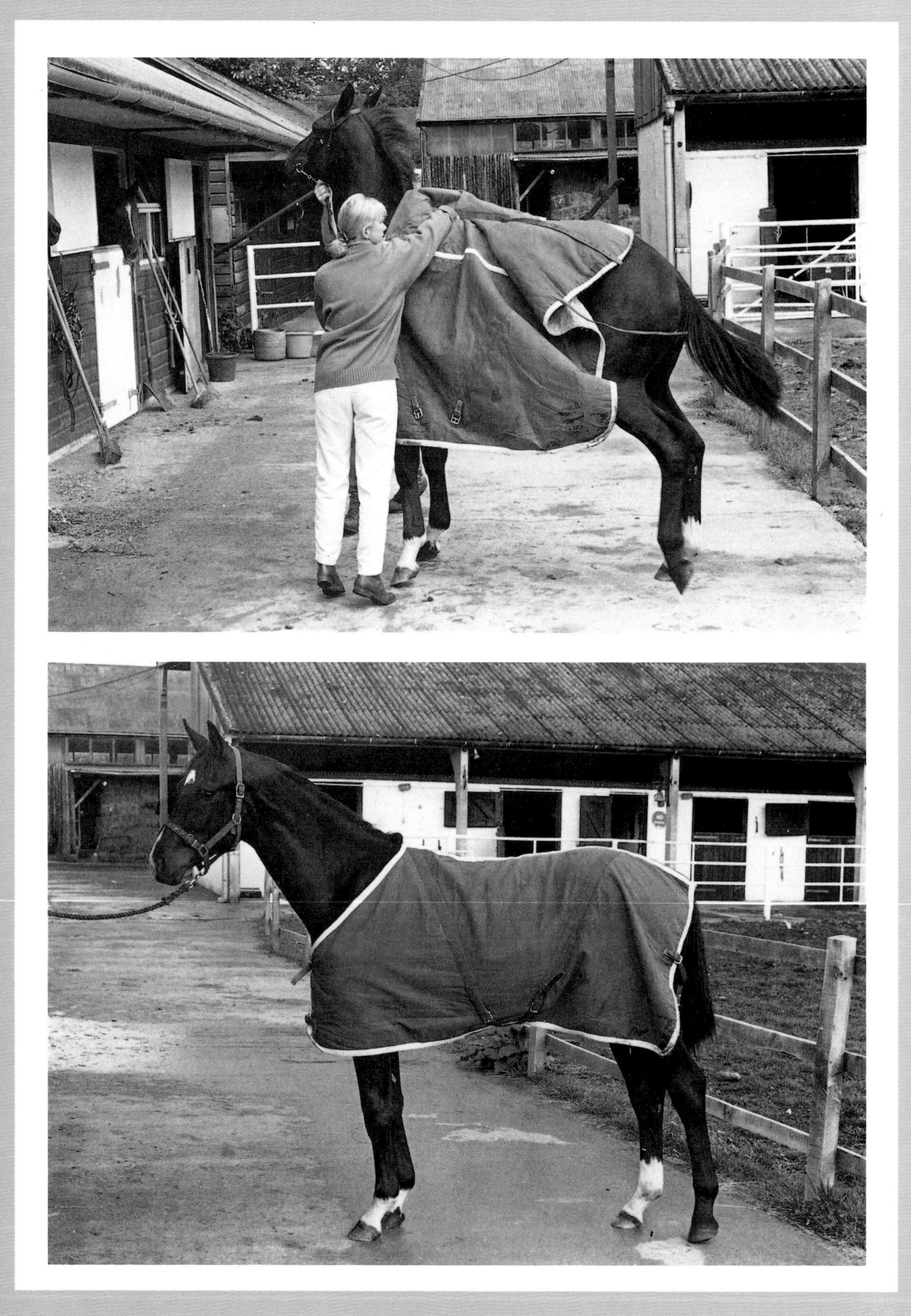

11 months: *AB being taught to stand up square. Do not worry if a yearling appears a little lightweight prior to being put out on to spring grass – he will soon change for the better!*

(Opposite above) *He may find the introduction to a rug somewhat daunting!*

(Opposite below) *A yearling is likely to try and destroy his first rug, so it is advisable to use one made of a hardwearing material such as canvas. Ensure that there are* no loose straps, *as these could be dangerous if he plays with his friends at grass*

14 months: *Training in hand is recommended. A colt of this age will need to be bitted; use a mild bit such as the loose-ring nylon straight-bar bit illustrated. If leading a youngster in a busy area it would be sensible to place yourself between the horse and the traffic*

(Opposite above) *AB being taught to trot up in hand, using a headcollar*
(Opposite below) 12 months: *It is time to have a colt castrated, or to separate him from his filly friends if he is to remain entire. However, companionship is most important to the youngster's well-being. AB is seen here with a gelding as companion*

E397 MDU
ALDERBOURNE LANE

22 months: *The end of the AB's second winter. Note how tall he is growing. His winter coat is still apparent. It is useful to practise occasionally to ensure he remembers his good manners*

(Opposite) 14 months: *AB's first introduction to the outside world: observing traffic, and a quiet walk down the lane. A good preparation for in-hand showing Note he is wearing a roller with side-reins loosely adjusted*

2 years old: *AB stood up: he is not yet 'full grown', but from his physique one can see how he is going to turn out as a mature horse*

(Opposite above) 2 years: *Early spring. AB's first lesson in lungeing*

(Opposite below) 2 years: *One month later. The lunge work is well established and will not be practised again during the summer of his second year*

APPENDIX I

GESTATION TABLE

DATE OF SERVICE	DATE OF BIRTH	DATE OF SERVICE	DATE OF BIRTH	DATE OF SERVICE	DATE OF BIRTH	DATE OF SERVICE	DATE OF BIRTH	DATE OF SERVICE	DATE OF BIRTH	DATE OF SERVICE	DATE OF BIRTH
MAR	FEB	APR	MAR	MAY	APR	JUNE	MAY	JULY	JUNE	AUG	JULY
1	3	1	6	1	5	1	6	1	5	1	6
2	4	2	7	2	6	2	7	2	6	2	7
3	5	3	8	3	7	3	8	3	7	3	8
4	6	4	9	4	8	4	9	4	8	4	9
5	7	5	10	5	9	5	10	5	9	5	10
6	8	6	11	6	10	6	11	6	10	6	11
7	9	7	12	7	11	7	12	7	11	7	12
8	10	8	13	8	12	8	13	8	12	8	13
9	11	9	14	9	13	9	14	9	13	9	14
10	12	10	15	10	14	10	15	10	14	10	15
11	13	11	16	11	15	11	16	11	15	11	16
12	14	12	17	12	16	12	17	12	16	12	17
13	15	13	18	13	17	13	18	13	17	13	18
14	16	14	19	14	18	14	19	14	18	14	19
15	17	15	20	15	19	15	20	15	19	15	20
16	18	16	21	16	20	16	21	16	20	16	21
17	19	17	22	17	21	17	22	17	21	17	22
18	20	18	23	18	21	18	23	18	22	18	23
19	21	19	24	19	23	19	24	19	23	19	24
20	22	20	25	20	24	20	25	20	24	20	25
21	23	21	26	21	25	21	26	21	25	21	26
22	24	22	27	22	26	22	27	22	26	22	27
23	25	23	28	23	27	23	28	23	27	23	28
24	26	24	29	24	28	24	29	24	28	24	29
25	27	25	30	25	29	25	30	25	29	25	30
26	28	26	31	26	30	26	31	26	30	26	31
MAR	MAR	APR	APR	MAY	MAY	JUNE	JUNE	JULY	JULY	AUG	AUG
27	1	27	1	27	1	27	1	27	1	27	1
28	2	28	2	28	2	28	2	28	2	28	2
29	3	29	3	29	3	29	3	29	3	29	3
30	4	30	4	30	4	30	4	30	4	30	4
31	5			31	5			31	5	31	5

APPENDIX II

EQUINE VIRAL ARTERITIS (EVA)

EVA was first diagnosed as a disease in the equine species in 1953. Before this time it had been included under the term 'influenza', and was known as 'pink eye'. EVA is found in nearly every country and is prevalent in Europe and North America.

The UK and Ireland have been well protected from the disease, partly due to the fact that they are islands and partly as a result of strict import regulations. The first EVA outbreak reported in the UK was in the summer of 1993. It was successfully controlled by immediate closing of infected and 'at risk' yards for a temporary period. However, professionals in the equine industry have little doubt that the disease will return in the foreseeable future.

Due to the fact that the equine population in these islands has no natural immunity to the disease it is likely to spread very rapidly. Such an occurrence could have a devastating effect on the racing industry and the non-Thoroughbred breeding programme.

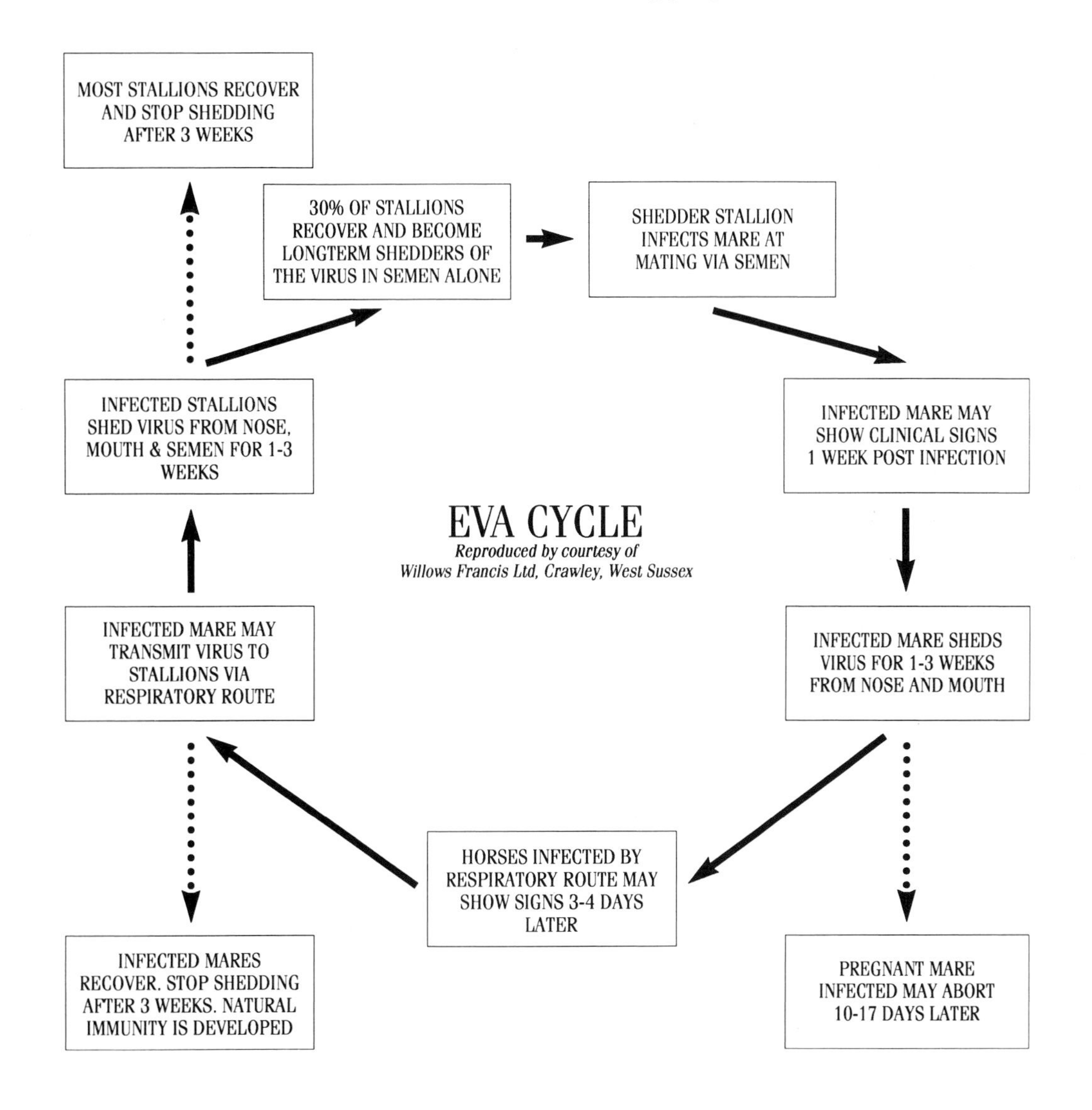

Reproduced by courtesy of Willows Francis Ltd, Crawley, West Sussex

Clinical signs

These are variable. The most common is fever, loss of appetite, nasal discharge, reddening of the membranes of the eye, swelling of the hind legs and depression. Sometimes the infection is not apparent and may even be fatal.

Mares may show swelling of the mammary glands and pregnant mares are likely to abort (60 per cent have been reported). Foals at foot may become affected, and can contact pneumonia as a result.

Stallions develop swelling of the scrotum which affects the quality of the semen.

Diagnosis

Due to the variability of clinical signs a laboratory diagnosis is necessary (swabs and/or blood samples).

Transmission

There are two ways in which EVA can be passed on:

1 In the acute stage of infection the virus is spread by direct contact during coughing, snorting or nasal secretion.

2 By the infected stallion who sheds virus in the semen (30 per cent of stallions who contact EVA are found to become permanent shedders and do not show any clinical signs).

Treatment

There is no specific treatment for EVA. The symptoms present will be treated by the veterinary surgeon as appropriate.

The control of EVA

All individual owners as well as breeding establishments should be vigilant in their observation for the signs of the disease.

All stallions and teasers should be serologically tested prior to the breeding season (including those used for AI).

All mares should be blood tested before mating in any breeding season.

Importation of stallions and mares Extreme care should be taken – consult your vet.

Vaccination There is a vaccination available – consult your vet for further advice.

NOTE **The Horserace Betting Levy Board** has published a Code of Practice for the control of EVA. To obtain copies please contact them at 52 Grosvenor Gardens, London SW1 0AU. Tel 071-333 0043.

FURTHER READING

Allen, W. Edward. *Fertility and Obstetrics in the Horse* (Blackwell Scientific Publications)

Ffrench Blake, R.L.V. *The Early Training of the Horse* (Seeley)

Hayes, Captain M. Horace. *Veterinary Notes for Horse Owners* (Stanley Paul)

Klimke, Reiner. *Basic Training of the Young Horse* (J. A. Allen)

Marrable, A. W. *The Foal in the Womb* (J. A. Allen)

Ripman, Barbara. *Basic Training* (Crowood)

Rose, John and Pilliner, Sarah. *Practical Stud Management* (Blackwell Scientific Publications)

Wynmalen, Henry. *Horse Breeding and Stud Management* (J. A. Allen)

INDEX